CASE STUDIES IN TRANSIENT HEAT TRANSFER WITH SENSITIVITIES TO GOVERNING VARIABLES

CASE STUDIES IN TRANSIENT HEAT TRANSFER WITH SENSITIVITIES TO GOVERNING VARIABLES

by
M. Kemal Atesmen

ASME PRESS

Library of Congress Cataloging-in-Publication Data

Names: Atesmen, M. Kemal, author.
Title: Case studies in transient heat transfer with sensitivities to governing variables / by M. Kemal Atesmen.
Description: New York, NY : ASME Press, 2023. | Includes bibliographical references and index. | Summary: "Time dependent energy conservation in a control volume refers to a fascinating and challenging portion of heat transfer called unsteady state or transient heat transfer. Energy conservation in first law of thermodynamics states that heat inflow into a control volume, plus the heat generated in a control volume, minus heat outflow from a control volume equals to change in internal energy of a control volume. Energy conservation differential equations for transient heat transfer in cartesian, cylindrical and spherical coordinates have been determined in numerous heat transfer text books such as References 10 and 15 and also presented in the Appendix of this book. This book investigates twenty-four transient heat transfer problems in our everyday lives. I chose wide variety of simple and every-day life problems in order to give readers insight into different approaches to a solution in transient heat transfer. Sensitivities of dependent variables to governing independent variables are investigated under appropriate physical conditions"-- Provided by publisher.
Identifiers: LCCN 2023021039 | ISBN 9780791886786 (hardback) | ISBN 9780791886793 (adobe pdf)
Subjects: LCSH: Heat--Transmission--Case studies. | Materials--Thermal properties--Case studies. | Thermal conductivity--Case studies. | Engineering mathematics--Case studies.
Classification: LCC TA418.54 .A467 2023 | DDC 621.402/20723--dc23/eng/20230506
LC record available at https://lccn.loc.gov/2023021039

Contents

Preface

Every heat transfer problem has to satisfy the first law of thermodynamics, namely the law of conservation of energy. Before steady-state conditions are reached, every heat transfer problem is in an unsteady state and energies in a control volume and energies coming in and out of a control volume can change with time. These time dependent energy conservation in a control volume refers to a fascinating and challenging part of heat transfer called the unsteady state or transient heat transfer. This book deals with twenty-four transient heat transfer problems in our everyday lives.

For example, our Sun's electromagnetic waves that reach our Earth make this place livable. Our Earth's spin provides us daily varying temperatures and its 23.5 degrees tilt with respect to its orbital plane provides us transient heat transfer all year around. Effects of our Sun on our Earth is modeled in this book by transient heat transfer in several chapters.

Nucleate pool boiling heat transfer rates are very high and therefore it is used to cool devices such as computers, nuclear reactors, rocket motors, etc. During nucleate pool boiling in a large superheated fluid, the bubble growth from a heated surface goes through several stages of transient heat transfer and it is detailed in this book.

During an annealing process, some solids such as spherical metal bearings with small radii, metal plates with small thicknesses and cylindrical shaped metals with small radii can lose or gain thermal energy through their surfaces with respect to time with little or no internal temperature gradients. Transient heat transfer problems in these cases are solved by the lumped capacitance method.

The surface boundary of a spacecraft entering Earth's atmosphere from space at very high speeds, about 18,000 mph, experiences very high heat loads. Ablating materials are used to capture most of the incoming heat loads to protect the spacecraft, its occupants and its instruments. Ablating materials have special characteristic such as low thermal conductivity, low density, low coefficient of thermal expansion, high emissivity and high melting temperatures. The partial differential equation for one-dimensional transient heat conduction in a semi-infinite body is solved to determine the thickness of the ablating material to capture more that 90% of the incoming heat load to protect the spacecraft, its occupants and its instruments.

In our daily cooking, we are always facing transient heat transfer issues such as temperature-time history of a hot pot of soup which is immersed in still air environment for cooling or temperature gradients through the white and the yoke of different size chicken eggs while assuming a spherically shaped egg which is suddenly immersed into a 100 C water surrounding it in order to hard boil it. We can investigate roasting a piece of lean beef which has a short cylindrical shape. Roasting is done in a convection oven at a certain temperature. The chef wants to know the time it takes to roast one piece beef as rare and another same size piece of beef as well done. Transient heat transfer can investigate how frost damage penetrates a spherical and ripe orange when there is a sudden change in outside temperature at night with convection heat transfer at the surface of the orange.

Time dependent heat transfer during solidification of magma in a lava lake in a pit crater is a natural phenomenon on our Earth. After volcanic eruption stops, magmas in these pit craters start to solidify and release latent heat of fusion to the atmosphere from its solidified surface. This magma solidification process can take years to complete. Time dependent heat transfer during solidification of magma in a lava lake in a pit crater is modeled and compared to actual solidified magma thickness versus time measurements.

The lumped capacitance method and analytical solutions cannot be applied to all transient and multi-dimensional heat transfer problems with complicated geometries and boundary conditions. I covered the explicit finite difference method application to a 2-dimensional transient heat transfer problem with multiple boundary conditions at surfaces of a stainless-steel rectangular prism bar and a triangular prism bar.

Transient temperature distributions in a large plane concrete wall which is subjected suddenly to convection heat transfer on both surfaces is analyzed using the explicit finite difference method. The only drawback in the explicit finite difference method is the stability in finite difference calculations. Same problem is also solved using the implicit finite difference method which does not require stability criteria in time increments.

Acknowledgements

Over 40 years of engineering, engineering management and project management in the global arena covering automotive, computer, data communication, and offshore oil industries were accomplished by exceptional support from my wife, Zeynep, and my family members. Some years I was away from home more than six months out of a year trying to tackle challenging engineering tasks.

I would like to dedicate this book to my excellent teachers and mentors in heat transfer and in fluid mechanics at several universities and organizations. Some of the names at the top of a long list are Prof. I. Flugge-Lotz, Prof. W. C. Reynolds, Prof. W. M. Kays, Prof. A. L. London, Prof. R. D. Haberstroh, Prof. L. V. Baldwin and Prof. T. N. Veziroglu.

I would like to dedicate this book to all engineering project teams' members with whom I had the pleasure of working together over the years, with enthusiasm, with imagination and with determination. Over these years my engineering project teams' members kept coming back to work along with me without any reservations.

M. Kemal Atesmen
Santa Barbara, California

About the Author

M. Kemal Atesmen completed his high school studies at Robert Academy in Istanbul Turkey in 1961. He received his B. Sc. degree from Case Western Reserve University, his M. Sc. degree from Stanford University, and his Ph. D. degree from Colorado State University, all in mechanical engineering. He is a life member of ASME. He initially pursued an academic and an industrial career in parallel and became an associate professor in mechanical engineering before dedicating his professional life to international engineering management and engineering project management for thirty-six years. He helped many young engineers in the international arena to bridge the gap between college and professional life in automotive, computer component, data communication, and offshore oil industries.

He published twelve books, sixteen technical papers, and has four patents. His books are "Global Engineering Project Management", CRC Press, 2008, "Everyday Heat Transfer Problems – Sensitivities to Governing Variables", ASME Press, 2009, "Understanding the World Around through Simple Mathematics", Infinity Publishing, 2011, "A Journey Through Life, Wilson Printing, 2013, "Project Management Case Studies and Lessons Learned", CRC Press, 2015, "Process Control Techniques for High Volume Production", CRC Press, 2016, "Engineering Management in a Global Environment: Guidelines and Procedures", CRC Press, 2017, "Case Studies in Fluid Mechanics with Sensitivities to Governing Variables", Wiley and ASME Press, 2018, "Nothing Is Fixed, Everything Changes", Archway Publishing, 2018, "Electromagnetic Waves and Heat Transfer: Sensitivities to Governing Variables in Everyday Life", ASME Press, 2021, "A Kid's Future = Excelling in Practical Mathematics Volume I: Pre-K through 6th Grade", Amazon Kindle Direct Publishing, 2021, "A Kid's Future = Excelling in Practical Mathematics Volume II: 7th Grade through 12th Grade", Amazon Kindle Direct Publishing, 2021.

Introduction

Time dependent energy conservation in a control volume refers to a fascinating and challenging portion of heat transfer called unsteady state or transient heat transfer. Energy conservation in first law of thermodynamics states that heat inflow into a control volume, plus the heat generated in a control volume, minus heat outflow from a control volume equals to change in internal energy of a control volume. Energy conservation differential equations for transient heat transfer in cartesian, cylindrical and spherical coordinates have been determined in numerous heat transfer text books such as References 10 and 15 and also presented in the Appendix of this book.

This book investigates twenty-four transient heat transfer problems in our everyday lives. I chose wide variety of simple and every-day life problems in order to give readers insight into different approaches to a solution in transient heat transfer. Sensitivities of dependent variables to governing independent variables are investigated under appropriate physical conditions. This book is an extension of my two other books in heat transfer. One of them covered generic problems in everyday heat transfer, Ref. [1], and the other one covered electromagnetic waves and heat transfer, Ref. [2]. This book is also an extension of another book of mine in fluid mechanics that covers case studies in fluid mechanics with sensitivities to governing variables, Ref. [3].

Simplest approach to a transient heat transfer problem is by using the lumped capacitance method where the resistance to conduction in any direction within a solid is much less than resistances to convection and to radiation heat transfer at its surfaces. In Chapters 2, 3, 5, 7, and 8, I provide solutions to different transient heat transfer problems using the lumped capacitance method with varying convection and radiation boundary conditions.

Next level of complexity in solving transient heat transfer problems is when all temperature gradients inside a body can only be in one dimension such as infinite plane walls, infinite cylinders, or spheres. For example, in a boiling spherical egg, the temperature gradient is assumed to be only in the radial direction. An exact solution to the transient and one-dimensional heat transfer equation in spherical coordinates helped me to solve the boiling spherical egg problem in Chapter 6 and frost damage penetration to a spherical and ripe orange in Chapter13.

An exact solution to the transient and one-dimensional heat transfer equation in cylindrical coordinates is used to heat treat pine logs with radiation heat transfer in Chapter 11. Exact solutions to the transient and one-dimensional heat transfer equation in rectangular coordinates with different boundary conditions are used in Chapters 1 and 4.

Semi-infinite solids have analytical solutions in transient heat transfer for different boundary conditions such as constant temperature, constant surface heat flux and constant surface convection. Semi-infinite solids' analytical solutions are used in Chapters 15, 16, 17, and 18.

It can be very complicated to determine soil temperatures with changing depth and time. Many factors such as soil composition and porosity, running, seeping or stationary water, ice, vegetation, trees, rain, snow, day and night temperature variations, global warming and human intervention, etc., affect soil temperature penetration. In Chapter 20, I investigated seasonally averaged varying temperature effects on a specific type of soil's depths using semi-infinite solid's analytical solution for a soil which is suddenly exposed to seasonally varying constant surface temperatures. I investigated in Chapter 21, transient infrared radiation heat transfer at night from a desert's surface which is suddenly exposed to outgoing and incoming constant infrared radiation heat flux using analytical solution to a semi-infinite desert surface. In Chapter 22, transient heat transfer from atmosphere's periodic surface temperature fluctuations to Earth's semi-infinite soil is analyzed. Temperature profiles versus soil depth in a seasonal cycle are determined.

When we get to analyze two and three-dimensional objects with complex boundary conditions in transient heat transfer, unsteady temperature distributions in these objects can only be determined by numerical finite difference methods such as the explicit method or the implicit method. Chapters 9 and 10 use explicit methods. Chapter 19 compares results of analytical solutions with both explicit and implicit finite difference methods.

Multidimensional objects can be expressed as products of one-dimensional infinite objects. A short cylindrical shape of a beef piece to be roasted can be analyzed as the product of an infinite cylinder and an infinite plate. Their product solutions are discussed in Chapter 12. For the three-dimensional transient temperature distribution of a rectangular parallelepiped, I used product of three one-dimensional exact solutions for three infinite plane walls, each in a different plane, in Chapter 14.

Analytical solution for transient heat transfer from an infinitely long and thin heat source, a thermal needle probe, which is used to measure thermal conductivity of soils and soft rocks in an infinite cylindrical medium, is treated in Chapter 23.

Transient heat transfer during solidification of magma in a lava lake in a pit crater is modeled by using a constant thermal conductivity for the solidified magma layer and results are compared with actual data in Chapter 24.

1

Bubble Growth in a Large Superheated Fluid During Nucleate Pool Boiling

Nucleate pool boiling heat transfer rates are very high and therefore it is used to cool devices such as computers, nuclear reactors, rocket motors, etc. During nucleate pool boiling in a large superheated fluid, the bubble growth from a heated surface goes through several stages. First bubble growing stage is mainly governed by surface tension and by inertia forces from the surrounding fluid. In low (T_{sup} - T_{sat}), i.e. (T_{sup} - T_{sat}) < 5 K, at atmospheric pressure, first bubble growth stage is less than 1 millisecond (ms). The bubble detaches from the heated surface when the buoyant force overcomes the surface tension and the bubble starts to grow close to a spherical shape. The second bubble growth stage is mainly governed by conduction and convection transient heat transfer. In low (T_{sup} - T_{sat}), the second bubble growing stage goes up to about 15 ms. Then, the growing bubble enters a third stage, where the fluid's growing thermal and velocity boundary layers around the bubble interact and govern the bubble growth. There have been many experimental and theoretical studies regarding bubble growth during nucleate pool boiling. An excellent overview of these studies is provided in Ref. [13]. In this chapter, we will model the second growing stage of the bubble surface in an infinite superheated fluid as a semi-infinite solid surface where only transient and one-dimensional heat conduction occurs as shown in Figure 1-1. This analysis assumes water as fluid interfacing a bubble at atmospheric pressure at low superheats, (T_{sup} - T_{sat}) < 5 K.

If a sudden change of temperature is imposed at the semi-infinite water bubble interface, transient and one-dimensional conduction heat transfer will apply by neglecting the momentum of superheated water at the water-bubble interface.

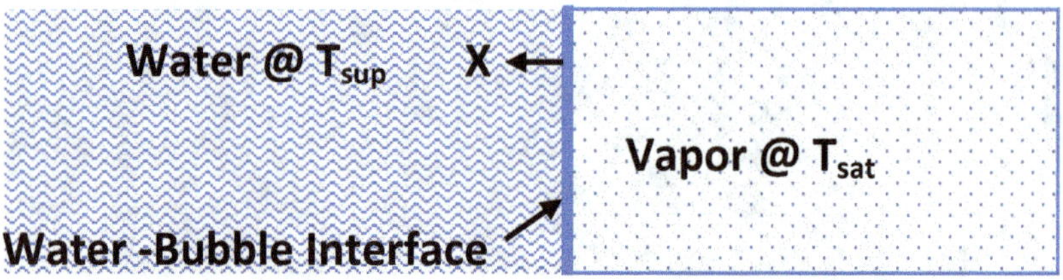

Figure 1-1 Semi-Infinite Solid Bubble Interface with Water

The partial differential equation that applies to this case with two independent variables, X and Θ, and with constant thermal diffusivity for water, α_w in m²/s, is given in Equation 1-1.

$$\frac{\partial T}{\partial \theta} = \alpha_w \frac{\partial^2 T}{\partial X^2} \qquad 1\text{-}1$$

where T is the temperature of water in K, Θ is time after sudden contact at the semi-infinite water-bubble interface and X is the horizontal distance in water's direction from the semi-infinite water-bubble interface. The initial condition that is used to solve Equation 1-1 is given in Equation 1-2.

$$T(X,0) = T_{sup} \qquad 1\text{-}2$$

Two boundary conditions that are used to solve Equation 1-1 are given in Equations 1-3.

$$T(0,\theta) = T_{sat} \quad \text{and} \quad T(\infty,\theta) = T_{sup} \qquad 1\text{-}3$$

An exact solution to the partial differential equation in Equation 1-1 subject to conditions in Equations 1-2 and 1-3 can be obtained by introducing a new similarity variable as shown in Equation 1-4.

$$U = \frac{X}{2\sqrt{\alpha_w \theta}} \qquad 1\text{-}4$$

The partial differential equation in Equation 1-1 becomes an ordinary differential as shown in Equation 1-5.

$$\frac{d^2 T}{dU^2} = -2U \frac{dT}{dU} \qquad 1\text{-}5$$

Boundary conditions in Equation 1-3 becomes as follows.

$$T(0) = T_{sat} \quad \text{and} \quad T(\infty) = T_{sup} \qquad 1\text{-}6$$

After separating variables in Equation 1-5, integrating twice and applying boundary conditions from Equation 1-6 provides the following temperature function in Equation 1-7 as X and Θ independent variables for a semi-infinite water bubble interface.

$$\frac{T-T_{sat}}{T_{sup}-T_{sat}}=\frac{2}{\sqrt{\pi}}\int_0^{\frac{X}{2\sqrt{a_w\theta}}}\exp\left(-m^2\right)\times dm=\operatorname{erf}\left(\frac{X}{2\sqrt{a_w\theta}}\right) \quad 1\text{-}7$$

where m is a dummy variable and $\operatorname{erf}\left(\frac{X}{2\sqrt{a_w\theta}}\right)$ is the Gaussian error function.

From equation 1-7, heat that is transferred at the semi-infinite water bubble interface can be obtained and it is given in Equation 1-8.

$$Q_{interface}=k_w\frac{\partial T(0,\theta)}{\partial X}=k_w\frac{T_{sup}-T_{sat}}{\sqrt{\pi a_w\theta}} \quad 1\text{-}8$$

where k_w is superheated water's thermal conductivity in W/(m-K). Now, the heat transferred to the bubble from the water interface can be related to the bubble radius growth by the following energy balance in Equation 1-9.

$$Q_{interface}=k_w\frac{T_{sup}-T_{sat}}{\sqrt{\pi a_w\theta}}=\rho_v H_{wv}\frac{dR}{d\theta} \quad 1\text{-}9$$

where ρ_v is the water vapor density in kg/m^3 and H_{wv} is the heat of vaporization of water in (W-s)/kg both at atmospheric pressure and at T_{sat}. R is the radius of the bubble in m. Equation 1-9 can be integrated and after the initial condition of $R = 0$ at $\theta = 0$ is applied, the following bubble-radius growth relationship can be obtained as shown in Equation 1-10.

$$R=\left(\frac{2}{\sqrt{\pi}}\right)\frac{k_w\left(T_{sup}-T_{sat}\right)}{\rho_v H_{wv}\sqrt{a_w}}\sqrt{\theta} \quad 1\text{-}10$$

Now, let us define the dimensionless Jacob number, Ja, as given in Equation 1-11. The Jacob number is defined for liquid-vapor phase change as the ratio of sensible heat energy carried by a liquid to the latent heat energy of a bubble with the same volume.

$$Ja=\frac{\rho_w C_{pw}\left(T_{sup}-T_{sat}\right)}{\rho_v H_{wv}} \quad 1\text{-}11$$

Bubble radius growth relationship can be rewritten using the Jacob number definition as shown in Equation 1-12.

$$R=\left(\frac{2}{\sqrt{\pi}}\right)Ja\sqrt{a_w\theta} \quad 1\text{-}12$$

Bubbles growing in an unbounded and superheated fluid are governed by complex interaction of heat transfer and fluid dynamics surrounding the bubble.

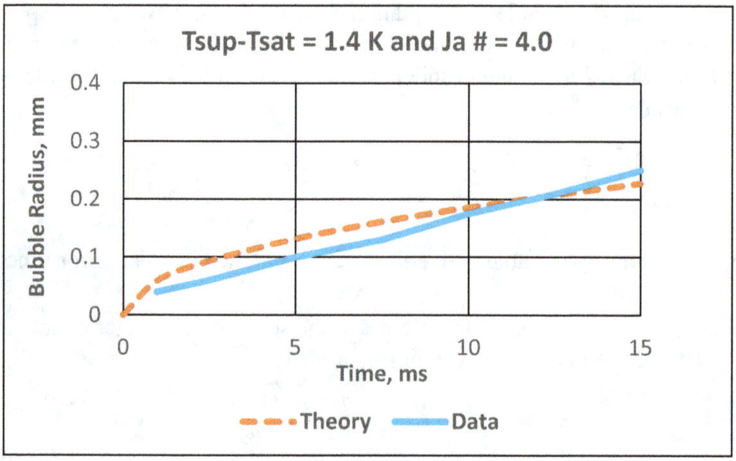

Figure 1-2 Bubble radius versus growth time for T_{sup}-T_{sat}=1.4 K, Ja # = 4.0

At low superheat temperature, i.e. $T_{sup} < 378\,K$, and at atmospheric pressure, radius of a growing bubble can be estimated by $R \propto \sqrt{\theta}$ when growth time is between 1 ms and 15 ms,

Next, we will use bubble growth experiments from Ref. [6] to test the above model which is for the second growing stage of a bubble's surface in an infinite superheated water at atmospheric pressure. The present model considers a semi-infinite solid interface between water and a bubble where only transient and one-dimensional heat conduction occurs. Bubble radius versus growth time for T_{sup}-T_{sat}=1.4 K, Ja # = 4.0 is shown in Figure 1.2.

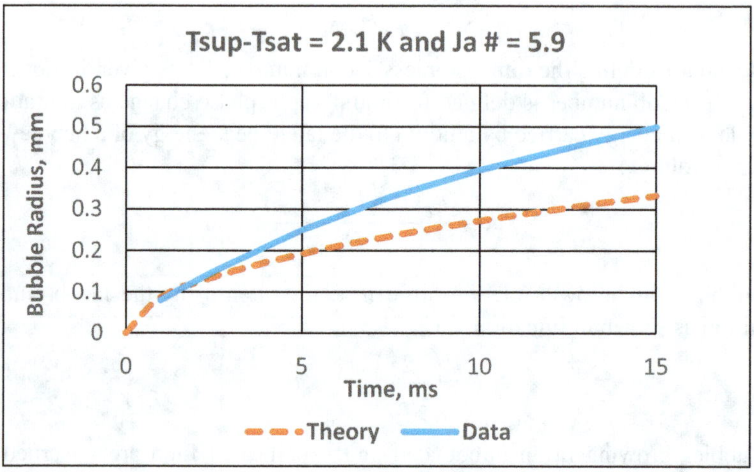

Figure 1-3 Bubble radius versus growth time for T_{sup}-T_{sat}=2.1 K, Ja # = 5.9

Bubble radius grows linearly up to 15 ms. Present model grows as square-root of time. The present model predicts the bubble radius growth adequately at water's low superheat condition, i.e. T_{sup}-T_{sat}=1.4 K.

In Figure 1-3, with a higher superheat condition of T_{sup} - T_{sat} = 2.1 K, experimental data for bubble radius grows as $\Theta^{0.68}$ up to 15 ms. Using the present model, however, bubble radius grows as square-root of time with this higher superheat condition. Bubbles growing in an unbounded and higher superheated water are governed by a more complex interaction of heat transfer and fluid dynamics surrounding the bubble's surface after couple of milliseconds of growth time.

2

Transient Heat Transfer with Little or No Temperature Gradient within Special Solids

Some solids such as spherical metal bearings with small radii, metal plates with small thicknesses and cylindrical shaped metals with small radii can lose or gain thermal energy through their surfaces with respect to time with little or no internal temperature gradients. Internal thermal energies of these special solids change with respect to time by heat gain or loss through their surfaces as given in Equation 2-1.

$$\rho_s V_s c_{p-s} \frac{dT_s}{d\theta} = -h_{Total} A_s \left(T_s - T_\infty \right) \qquad \text{2-1}$$

where ρ_s is the density of the solid piece in kg/m^3, V_s is the volume of the solid piece in m^3, c_{p-s} is the specific heat of the solid piece at constant pressure in (W-s)/(kg-K), T_s is the temperature of the solid piece in K, θ is time in seconds, h_{Total} is the sum of radiation and convection heat transfer coefficients at the surface of the solid piece in W/(m^2-K), A_s is the surface area of the solid piece in m^2 and T_∞ is the environment temperature surrounding the solid piece in K.

For the transient thermal energy balance to hold as given in Equation 2-1, the thermal conduction resistance of the solid piece, $\frac{L_{ch-s}}{k_s}$, has to be less than 10% of total thermal surface resistances of the solid piece due to convection and radiation, namely, $\frac{L_{ch-s}}{k_s} < 0.1\left(\frac{1}{h_{Total}}\right)$. $L_{ch-s} = \frac{V_s}{A_s}$ is called the characteristic length of the solid piece in meters which is the ratio of its volume to its surface area. k_s is the thermal conductivity of the solid piece in W/(m-K).

In Equation 2-1, it is assumed that both radiation and convection heat transfers are occurring through same surfaces of the solid piece, namely through surfaces A_s. In most cases, during heating and cooling of a solid piece, h_{total} is the total heat transfer coefficient at solid's surfaces and it is a function of time as shown in Equation 2-2.

$$h_{Total}(\theta) = h_{Radiation}(\theta) + h_{Convection}(\theta) \qquad 2\text{-}2$$

The validity of Equation 2-1 is defined by a dimensionless Biot number which is the ratio of the internal thermal conduction resistance to the total surface thermal radiation and convection resistances of a solid piece as shown in Equation 2-3.

$$Bi = \frac{L_{ch-s}/k_s}{1/h_{Total}} = \frac{h_{Total}L_{ch-s}}{k_s} < 0.1 \qquad 2\text{-}3$$

Combination of Equations 2-1 and 2-3 defines the simple lumped capacitance analysis method which is very important in solving transient heat transfer problems. $\left(\rho_s V_s c_{p-s}\right)$ is the lumped thermal capacitance of the solid and $\left(\frac{1}{h_{Total}A_s}\right)$ is the total resistance to convection and to radiation at solid's boundary. Equation 2-3 verifies that temperature gradients within the solid are negligible.

Let us now investigate heating and cooling of a square stainless-steel plate with a width, W, $W = 2\ m$, a length, L, $L = 2\ m$, and a thickness, t, $t = 0.01\ m$, during a three-stage annealing process for stress relieving. The stainless-steel plate is heated from 25 C to 600 C in the first stage annealing oven which has a surrounding temperature, both oven air and walls, at 800 C. After the stainless-steel plate reaches 600 C, it is transferred to a second stage oven where the oven air and walls are at 600 C and the plate stays there for 20 minutes. Then, the stainless-steel plate is transferred to a third stage oven for cooling where the oven air and walls are at 25 C. The stainless-steel plate cools down to 50 C in this third oven's environment and then it is taken out of the annealing process.

The characteristic length of this stainless-steel plate is shown in Equation 2-4.

$$L_{ch-s} = \frac{V_s}{A_s} = \frac{WLt}{2WL\left(1+\frac{t}{L}+\frac{t}{W}\right)} \approx \frac{t}{2} = \frac{0.01}{2} = 0.005\,m. \qquad 2.4$$

The radiation heat transfer coefficient, $h_{Radiation}$, during heating and cooling stages of the annealing process are given in Equation 2-5.

$$h_{Radiation} = \frac{\varepsilon_s\sigma\left(T_s^4 - T_\infty^4\right)}{\left(T_s - T_\infty\right)} = \varepsilon_s\sigma\left(T_s + T_\infty\right)\left(T_s^2 + T_\infty^2\right) \qquad 2\text{-}5$$

where ε_s is the hemispherical emissivity of stainless-steel plate's surfaces and σ is the Stephan Boltzmann constant at $5.67\text{x}10^{-8}$ W/(m^2-K^4).

The free convection heat transfer coefficient, $h_{Convection}$, during heating and cooling between vertical stainless-steel plate's surfaces and oven air is given by the following empirical Equation 2-6, see Ref. [10].

$$\overline{Nu_L} = \frac{\overline{h}_{Convection}}{\left(k_{air}/L\right)} = \left\{0.825 + \frac{0.387 Ra_L^{1/6}}{\left[1+\left(0.492/Pr\right)^{9/16}\right]^{8/27}}\right\}^2 \qquad 2\text{-}6$$

where, $Ra_L = \dfrac{g\beta\left(T_s - T_\infty\right)L^3}{\nu a}$, is the Rayleigh number which is the ratio of buoyancy to viscous forces times the ratio of momentum to thermal diffusivities, namely the Prandtl number, for oven air. $L = W = 2\ m$ is the length or width of the vertical and square stainless-steel plate. k_{air} is the thermal conductivity of the oven air in W/(m-K). g is the gravitational constant at sea level, namely 9.81 m/s^2.

Oven air is assumed to be an ideal gas and its effective temperature is the film temperature, $T_{Film} = 0.5\left(T_s + T_\infty\right)$. All oven air properties are calculated at T_{Film} temperature. $\beta = \dfrac{1}{T_{Film}}$ is the volumetric thermal expansion coefficient for oven air in 1/K. ν is oven air's momentum diffusivity in m^2/s. $a = \dfrac{k_{air}}{\rho c_p}$ is oven air's thermal diffusivity in m^2/s, ρ is density of oven air in kg/m^3, and c_p is the specific heat of oven air at constant pressure in (W-s)/(kg-K). The Prandtl number, $Pr = \dfrac{\nu}{a}$, is the ratio of momentum and thermal diffusivities for oven air.

Equation 2-1 is solved numerically by using forward Euler method which is an explicit numerical method where $T_{s,j+1}$, stainless steel plate's temperature after a time increment $\Delta\theta$ seconds is determined from values that are already known such as $T_{s,j}$, $T_{Film,j}$, $h_{Radiation}$, and $h_{Convection,j}$. Equation 2-1 is rewritten in Equation 2-7 for explicit numerical method.

$$T_{s,j+1} = T_{s,j} - \left(\frac{\Delta\theta}{\rho_s c_{p-s} L_{ch-s}}\right)\left(h_{Radiation,j} + h_{Convection,j}\right)\left(T_{s,j} - T_\infty\right) \qquad 2\text{-}7$$

Equation 2-7 can be rewritten in a non-dimensional form by introducing one more dimensionless number, Fo, the Fourier number. The Fourier number is the dimensionless time step for the numerical calculations as shown in Equation 2-8.

$$Fo = \frac{a_s \Delta\theta}{L_{ch-s}^2} \qquad 2\text{-}8$$

Equation 2-7 takes the non-dimensional form shown in Equation 2-9.

$$\frac{T_{s,j+1} - T_{s,j}}{T_{s,j} - T_\infty} = BiFo \qquad 2.9$$

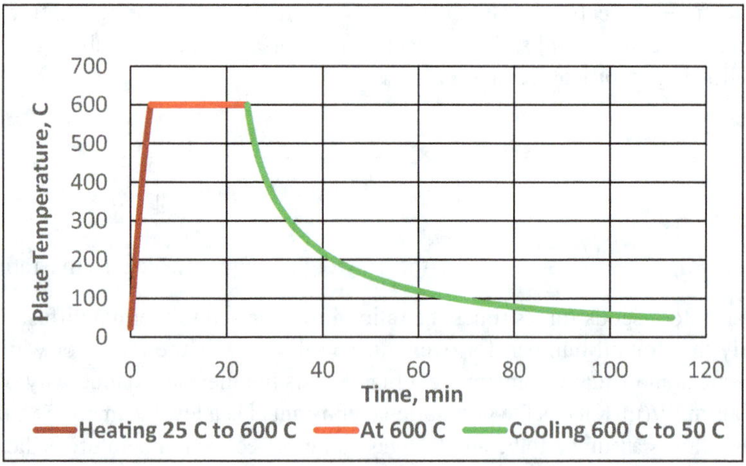

Figure 2-1 Stainless-steel temperature versus time for three annealing stages

The temperature profile for the stainless-steel plate in this three-stage annealing process is shown in Figure 2-1.

Heating time to raise the plate's temperature from 25 C to 600 C is 4.25 minutes in the first annealing stage oven with an 800 C environment. The 600 C soak is 20 minutes in the second stage annealing oven with a 600 C environment. Finally, the cooling time for the plate from 600 C down to 50 C is 88.58 minutes in the third stage oven with a 25 C environment. The total three stage annealing process for this stainless-steel took 112.83 minutes. The maximum total heat transfer coefficient during annealing occurs at the end of the heating stage at an 800 C environment when $h_{Total-max} = 119.84 W/(m^2 - \mathrm{K})$. Therefore, the maximum Biot number occurs at this point in time which is shown in Equation 2-10.

$$Bi_{max} = \frac{h_{Total-max} L_{ch-s}}{k_{ss}} = \frac{119.84\left(\frac{5}{1000}\right)}{48} = 0.0125 \qquad 2\text{-}10$$

For the present numerical calculations $\Delta\theta = 5\ s$. Smaller $\Delta\theta$ such as 1 s does not change results of the present annealing process. Heat transfer coefficients, total, radiation and free convection, during heating and cooling stages are shown in Figures 2-2 and 2-3.

During the heating stage of the annealing process, radiation heat transfer initially affects the temperature increase of the stainless-steel plate 89% and its effect increases to 96% as the plate plate's temperature approaches to 600 C and the effect of the free convection heat transfer decreases.

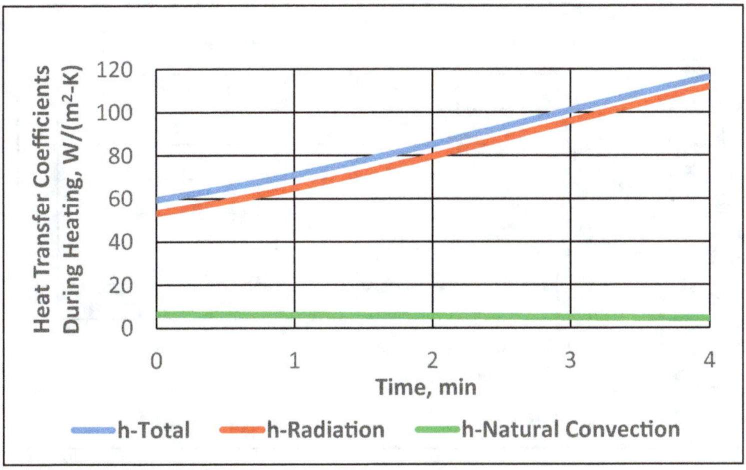

Figure 2-2 Heat transfer coefficients during the heating stage of annealing process

During the cooling stage of the annealing process, radiation heat transfer initially affects the temperature decrease of the stainless-steel plate 83% and its effect decreases to 63% as the plate plate's temperature approaches to 50 C and the effect of the free convection heat transfer approaches the effect of the radiation heat transfer. Free convection heat transfer initially affects the temperature decrease of the stainless-steel plate 17% and its effect increases to 37% as the plate plate's temperature approaches to 50 C.

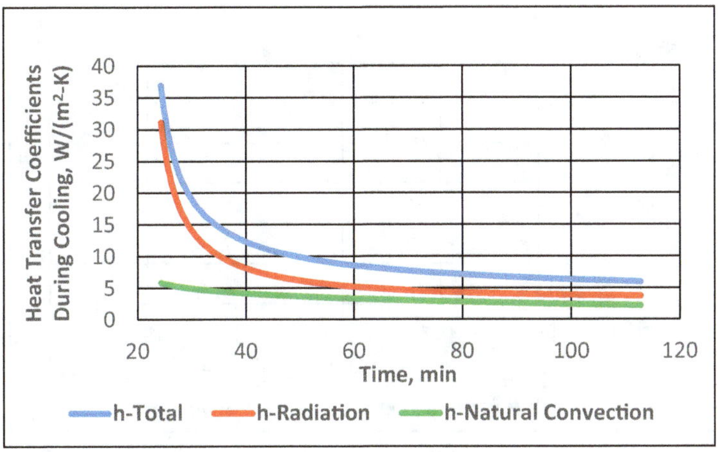

Figure 2-3 Heat transfer coefficients during the cooling stage of annealing process

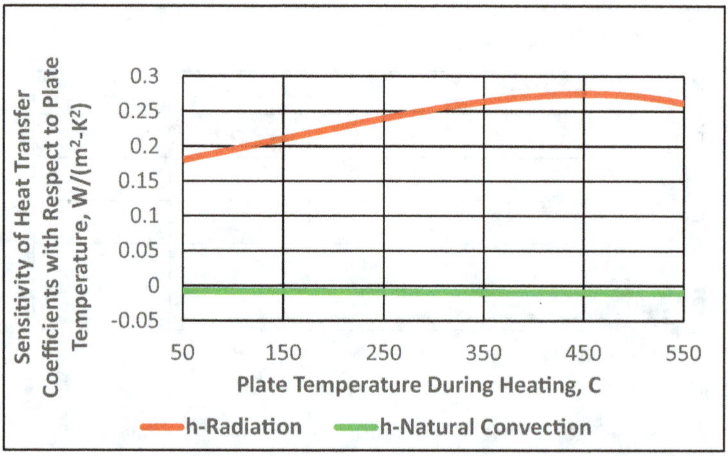

Figure 2-4 In Figure 2-4, as the plate's temperature increases Sensitivity of heat transfer coefficients to plate temperature versus plate temperature during heating in first stage annealing oven

Sensitivity of heat transfer coefficients to plate temperature versus plate temperature during heating in the first stage annealing oven is shown in Figure 2-4. As the plate's temperature increases from 25 C to 600 C, the radiation heat transfer coefficient's sensitivity to plate temperature increases, then the sensitivity goes through a maximum at the plate temperature of 450 C, and the sensitivity starts to decrease from 450 C to 600 C.

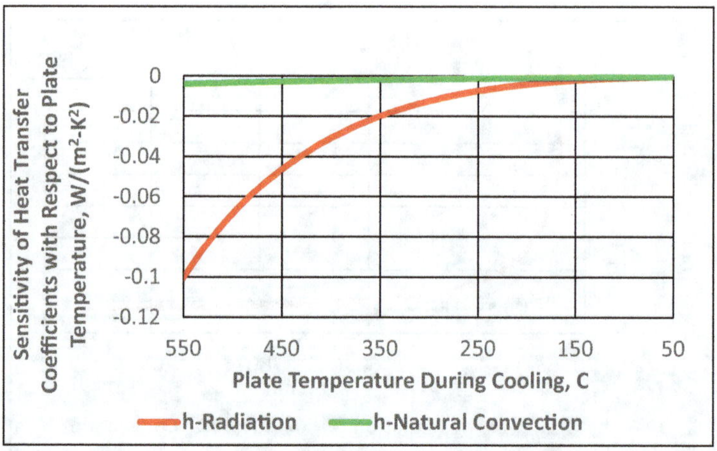

Figure 2-5 Sensitivity of heat transfer coefficients with respect to plate temperature versus plate temperature during cooling in third stage annealing oven

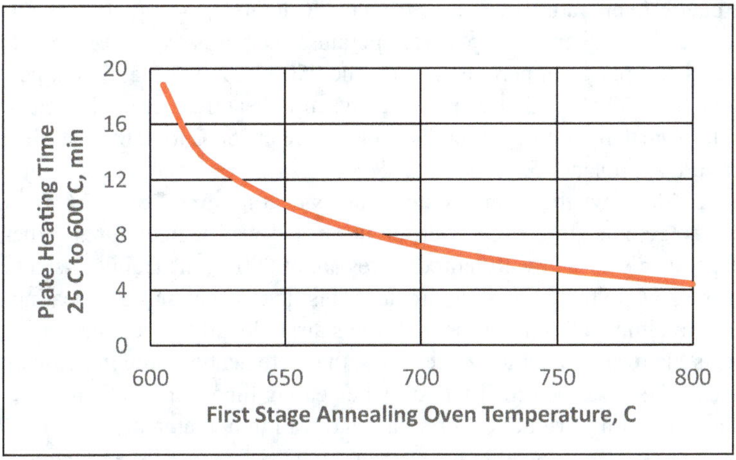

Figure 2-6 Plate heating time from 25 C to 600 C in a first stage annealing oven at different air and wall temperatures

In Figure 2-4, as the plate's temperature increases from 25 C to 600 C, the free convection heat transfer coefficient's sensitivity to plate temperature is always negative and it varies from -0.007 to -0.009.

Sensitivity of heat transfer coefficients to plate temperature versus plate temperature during cooling in the third stage annealing oven is shown in Figure 2-5.

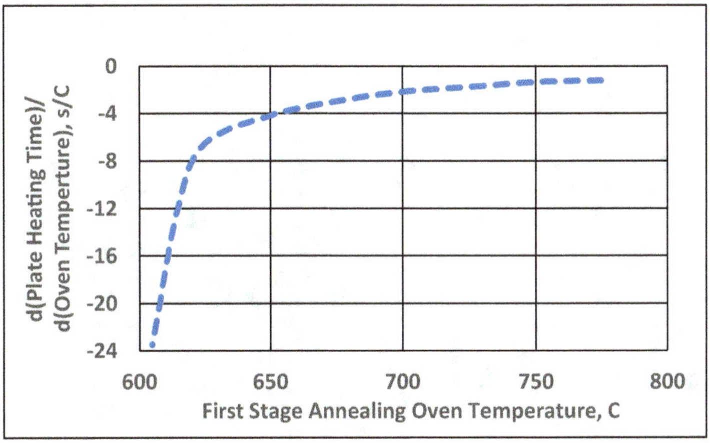

Figure 2-7 Sensitivity of plate heating time from 25 C to 600 C in a first stage annealing oven to different air and wall temperatures

As the plate's temperature decreases from 600 C to 50 C, the radiation heat transfer coefficient's sensitivity to plate temperature increases from -0.1 at 550 C to -0.0002 as the plate temperature approaches 50 C. As the plate's temperature decreases from 600 C to 50 C, the free convection heat transfer coefficient's sensitivity to plate temperature increases from -0.004 at 550 C to -0.0002 as the plate temperature approaches 50 C.

Figure 2-6 shows the stainless-steel plate's heating time from 25 C to 600 C in a first stage annealing oven at different air and wall temperatures. When the first stage oven's air and wall temperatures are at 800 C, the stainless-steel plate heats from 25 C to 600 C in 4.25 minutes. This first annealing stage heating time increases to almost 20 minutes as the oven's air and wall temperatures approach 605 C. As the oven temperature decreases, the plate heating time increases.

Figure 2-7 shows the sensitivity of plate heating time from 25 C to 600 C in a first stage annealing oven to different air and wall temperatures. About at 775 C oven temperature, plate heating time sensitivity is -1.25 seconds per degree centigrade. About at 650 C oven temperature, plate heating time sensitivity decreases to -4.2 seconds per degree centigrade. As the oven temperature approaches 600 C, the plate heating time increases with an absolute slope of 23.9 seconds per degree centigrade at about 605 C oven temperature.

3

Transient Heat Transfer Lumped Capacitance Method Applied to Composite Systems

The simple lumped capacitance method of solving transient heat transfer problems was introduced in Chapter 2. This method can also be applied to composite systems as long as each lump of the composite system satisfies $Bi_{lump} < 0.1$. In this chapter, we will determine the temperature-time history of water in a thin-walled cubic aluminum container which is immersed in engine oil which is also contained in a thin-walled chromium-steel container as shown in Figure 3-1. Initially, this whole composite system is at 20 C and it is suddenly immersed into a large environment that is filled with water at 90 C.

Each section in this composite system has substantial thermal capacitance, ($C_p\rho V$), where C_p is the specific heat in (W-s)/(kg-K), ρ is the density in kg/m^3.

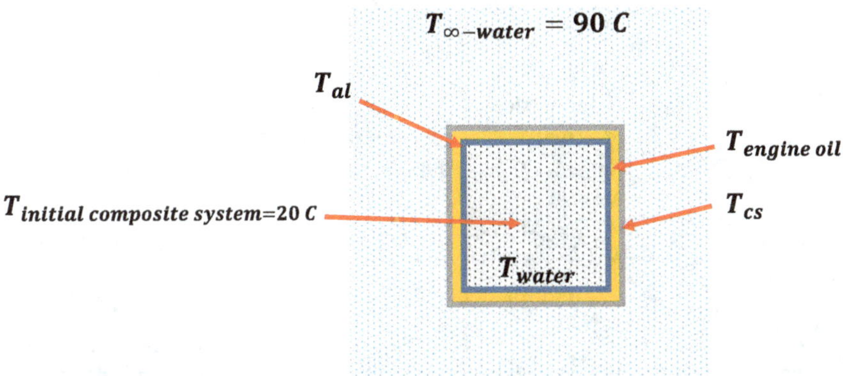

Figure 3-1 A cubic composite system of water-aluminum container-engine oil-chromium steel container initially at 20 C is suddenly immersed into a large environment of water at 90 C

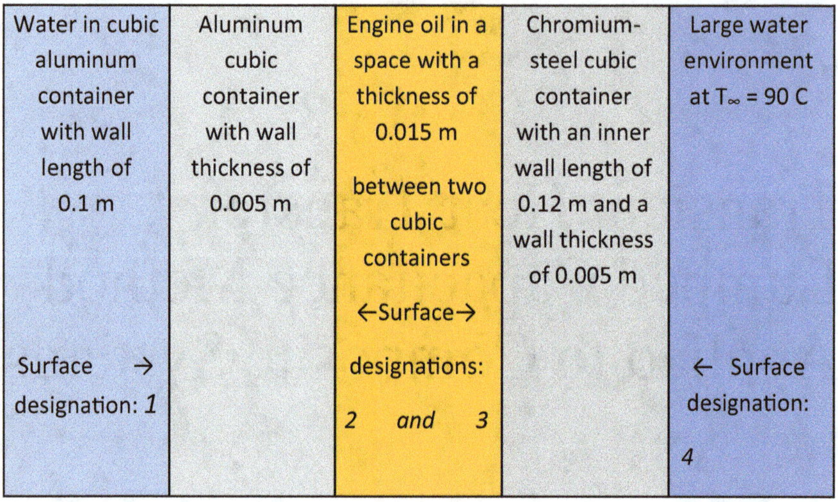

Figure 3-2 Composite system sections and surface designations

V is the volume in m^3 as shown in Figure 3-2 and Table 3-1. Each section of the composite system with thermal capacitance is called the lump of the system.

In the present analysis of temperature-time history for each section, properties are determined at the film temperature of $T_{film} = 0.5(20 + 90) = 55\ C$. Heat transfer coefficients on surfaces 1, 2, 3, and 4 are determined by using the empirical relationship for laminar free convection given in Equation 3-1. See Ref. [10].

$$\overline{Nu_L} = 0.59 \times Ra_L^{0.25} \qquad 3\text{-}1$$

Sections	Volumetric Heat Capacity C_p, (W-s)/(kg-K)	Density ρ, kg/m^3	Volume V, m^3	$(C_p\rho V)$, (W-s)/K
Water	4,184	984	$1.00\text{x}10^{-3}$	$C_w = 4{,}118$
Aluminum Container	875	2,770	$1.58\text{x}10^{-4}$	$C_{al} = 382$
Engine Oil	2,035	866	$5.70\text{x}10^{-4}$	$C_{eoil} = 1{,}005$
Chromium-steel Container	443	7,836	$2.25\text{x}10^{-4}$	$C_{cs} = 781$

Table 3-1 Thermal capacitance for each section (lump) of the composite system

where $\overline{Nu_L} = \dfrac{\left(\dfrac{L}{k_{fluid}}\right)}{\left(\dfrac{1}{h}\right)}$ is the Nusselt number which represents the dimensionless temperature gradient at the fluid-metal surface. $Ra_L = \dfrac{g\beta\left(T_\infty - T_{film}\right)L^3}{\nu\alpha}$ is the Rayleigh number which is the product of Grashof number (ratio of buoyancy to viscous forces) and Prandtl number (ratio of momentum and thermal diffusivities). g is the gravitational acceleration at sea level in m/s². β is the volumetric thermal expansion coefficient of fluid in 1/K. $T_\infty = 363\,K$. $T_{film} = 328\,K$. L is the characteristic length for the free convection heat transfer surface in m. ν is the kinematic viscosity of fluid in m²/s. α is the thermal diffusivity of fluid in m²/s. Average laminar free convection heat transfer coefficients for four surfaces of the composite system are given in Table 3-2.

Thermal resistance at each fluid surface is presented in Table 3-3.

Average laminar heat transfer coefficients, W/(m² - K)	
$\overline{h_1}$, Water and aluminum container's inner surfaces	637
$\overline{h_2}$, Engine oil and aluminum container's outer surfaces	341
$\overline{h_3}$, Engine oil and chromium-steel container's inner surfaces	333
$\overline{h_4}$, Large water environment and chromium-steel container's outer surfaces	988

Table 3-2 Average laminar heat transfer coefficients obtained using Equation 3-1

Average thermal resistance of each surface, K/W	
$R_1 = 1/(\overline{h}_1 \times A_1)$, Water and aluminum container's inner surfaces	0.0262
$R_2 = 1/(\overline{h}_2 \times A_2)$, Engine oil and aluminum container's outer surfaces	0.0444
$R_3 = 1/(\overline{h}_3 \times A_3)$, Engine oil and chromium-steel container's inner surfaces	0.0347
$R_4 = 1/(\overline{h}_4 \times A_4)$, Large water environment and chromium-steel container's outer surfaces	0.0108

Table 3-3 Average thermal resistance for each surface of the composite system

Here we have a composite system with four lumped thermal capacitances. We will assume uniform temperature distributions through walls of the aluminum and the chromium-steel containers by checking the ratios of conduction resistances in solid materials with respect to convection resistances at their surfaces, namely the Biot number, $Bi = \frac{L_{ch-solid} / k_{solid}}{(1/h)_{surface}}$. In order to be able to neglect temperature gradients in solid materials, Equation 3-2 has to be satisfied.

$$Bi = \frac{L_{ch-solid} / k_{solid}}{(1/h)_{surface}} < 0.1 \qquad \text{3-2}$$

Biot numbers for the present composite system's solids are shown in Table 3-4 with a characteristic length of L_{ch}= 0.5 × 0.005=0.0025 *m.* for both solids.

Biot numbers for solid's inner and outer surfaces	
$Bi_{1-al} = \left(\frac{L_{cl}}{k_{al}}\right) / \left(\frac{1}{h_1}\right)$, Aluminum container's inner surfaces with water	0.0088
$Bi_{2-al} = \left(\frac{L_{cl}}{k_{al}}\right) / \left(\frac{1}{h_2}\right)$, Aluminum container's outer surfaces with engine oil	0.0047
$Bi_{3-cs} = \left(\frac{L_{cl}}{k_{cs}}\right) / \left(\frac{1}{h_3}\right)$, Chromium-steel container's inner surfaces with engine oil	0.017
$Bi_{4-cs} = \left(\frac{L_{cl}}{k_{cs}}\right) / \left(\frac{1}{h_4}\right)$, Chromium-steel container's outer surfaces with large water environment	0.050

Table 3-4 Biot numbers for solid's inner and outer surfaces

With above assumptions, energy balances for each section or lump of the composite system for water (w), aluminum container (al), engine oil (eoil), and chromium-steel (cs) are shown below in Equations 3-3 through 3-8.

$$\left(C_p \rho V\right)_w \frac{dT_w}{d\theta} = h_1 A_1 \left(T_{al} - T_w\right) \tag{3-3}$$

$$\left(C_p \rho V\right)_{al} \frac{dT_{al}}{d\theta} = h_2 A_2 \left(T_{eoil} - T_{al}\right) - h_1 A_1 \left(T_{al} - T_w\right) \tag{3-4}$$

$$\left(C_p \rho V\right)_{eoil} \frac{dT_{eoil}}{d\theta} = h_3 A_3 \left(T_{cs} - T_{eoil}\right) - h_2 A_2 \left(T_{eoil} - T_{al}\right) \tag{3-5}$$

$$\left(C_p \rho V\right)_{cs} \frac{dT_{cs}}{d\theta} = h_4 A_4 \left(T_{\infty} - T_{cs}\right) - h_3 A_3 \left(T_{cs} - T_{eoil}\right) \tag{3-6}$$

Initial temperature for each section: $T_w = T_{al} = T_{eoil} = T_{cs} = 20C$ 3-7

Large environment of water temperature: $T_{\infty} = 90C$ 3-8

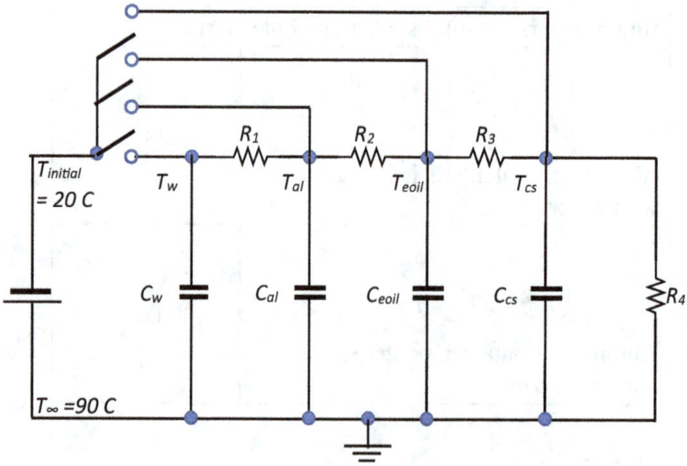

Figure 3-3 Thermal network analogy of the four-section or lump composite system

The thermal network analogy of the four-section or lump system is shown in Figure 3-3.

For the four-section or lump composite system, there are four coupled first order differential Equations 3-3 through 3-6 and four initial conditions in Equation 3-7. Equations 3-3 through 3-8 are solved numerically by using forward Euler method which is an explicit numerical method where $T_{w,j+1}$, $T_{al,j+1}$, $T_{eoil,j+1}$, and $T_{cs,j+1}$ temperatures with a time increment $\Delta\theta = 1s$ are determined from previous temperature values that are already known. All capacitance and resistance values are evaluated at the film temperature of 55 C and used as constants during numerical calculations. Equations 3-3 through 3-6 are rewritten in Equations 3-9 through 3-12 for explicit numerical method.

$$T_{w,j+1} = T_{w,j} + \frac{\overline{h}_1 A_1 \Delta\theta}{\left(C_p \rho V\right)_w}\left(T_{al,j} - T_{w,j}\right) \qquad 3\text{-}9$$

$$T_{al,j+1} = T_{al,j} + \frac{\overline{h}_2 A_2 \Delta\theta}{\left(C_p \rho V\right)_{al}}\left(T_{eoil,j} - T_{al,j}\right) - \frac{\overline{h}_1 A_1 \Delta\theta}{\left(C_p \rho V\right)_{al}}\left(T_{al,j} - T_{w,j}\right) \qquad 3\text{-}10$$

$$T_{eoil,j+1} = T_{eoil,j} + \frac{\overline{h}_3 A_3 \Delta\theta}{\left(C_p \rho V\right)_{eoil}}\left(T_{cs,j} - T_{eoil,j}\right) - \frac{\overline{h}_2 A_2 \Delta\theta}{\left(C_p \rho V\right)_{eoil}}\left(T_{eoil,j} - T_{al,j}\right) \qquad 3\text{-}11$$

$$T_{cs,j+1} = T_{cs,j} + \frac{\overline{h}_4 A_4 \Delta\theta}{\left(C_p \rho V\right)_{cs}}\left(T_{\infty,j} - T_{cs,j}\right) - \frac{\overline{h}_3 A_3 \Delta\theta}{\left(C_p \rho V\right)_{cs}}\left(T_{cs,j} - T_{eoil,j}\right) \qquad 3\text{-}12$$

Temperature-time profiles for each section or lump for transient heat transfer for the present composite system with substantial thermal capacitances are shown in Figures 3-4 and 3-5. Water temperature (blue color) increases much slower due to high thermal capacitances in aluminum container walls, in engine oil layer and in chromium-steel container walls. Also, high resistances at surfaces R_2 and R_3 contribute to slow water temperature rise.

The chromium-steel container's outer walls have the lowest time constant in the composite system, $\tau_{4-cs} = R_4 C_{cs}$, and its temperature increases the fastest to 90 C as shown in Figures 3-4 and 3-5 and in Equation 3-13.

$$\tau_{4-cs} = R_4 C_{cs} = (0.0108)\frac{K}{W}(781)\frac{W-s}{K} = 8.43\,s \qquad \text{3-13}$$

The chromium-steel container's outer walls reach a temperature of 60.9 C in 8.43 s which is a 58% rise in a 70 C temperature potential. On the other hand, water in the aluminum container has the highest time constant in the composite system and its temperature increases the slowest to 90 C as shown in Figures 3-4 and 3-5 and in Equation 3-14.

$$\tau_{1-w} = R_1 C_w = (0.0262)\frac{K}{W}(4{,}118)\frac{W-s}{K} = 107.72\,s \qquad \text{3-14}$$

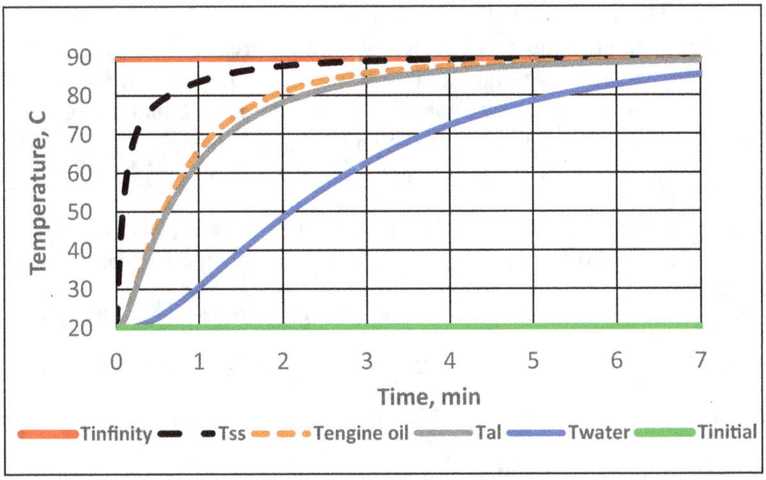

Figure 3-4 Temperature-time profiles for water in the aluminum container, for aluminum container walls, for engine oil, and for chromium-steel container walls from initial immersion at $\theta = 0$ *min* to $\theta = 7$ *min*

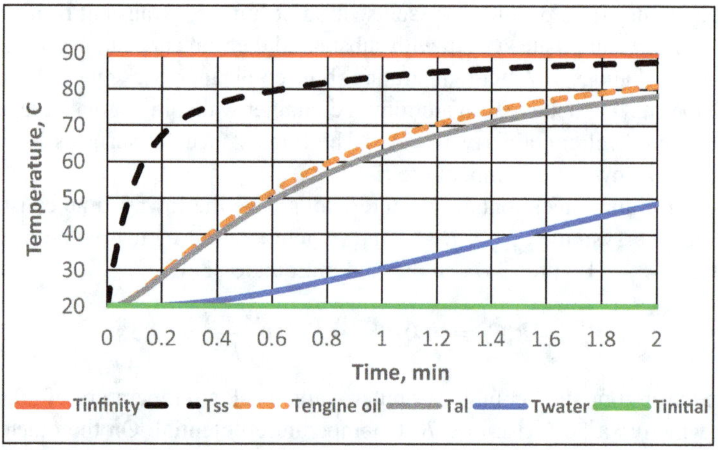

Figure 3-5 Temperature-time profiles for water in the aluminum container, for aluminum container walls, for engine oil, and for chromium-steel container walls from initial immersion at $\theta = 0\ min$ to $\theta = 2\ min$

Water in the aluminum container reaches a temperature of only 45 C in 107.72 s which is a 36% rise in a 70 C temperature potential.

From temperature-time profiles presented in Figure 3-4 for water in the aluminum container, for aluminum container walls, for engine oil, and for chromium-steel container walls, rate of change of internal thermal energy can be obtained for each section or lump as shown in Figure 3-6.

Rate of change of internal thermal energy, $\left(C_p\rho V\right)_{cs}\dfrac{dT_{cs}}{d\theta}$, in the chromium-steel container starts at 4,633 Watts at θ = 3 s and decrease fast down to 224 Watts after 30 s or 0.5 min. Rate of change of internal thermal energy, $\left(C_p\rho V\right)_{eoil}\dfrac{dT_{eoil}}{d\theta}$, in the engine oil starts at 434 Watts at $\theta = 3\ s$, goes through a maximum of 1096 Watts at $\theta = 14\,s\ or\ 0.23\,min$, and decreases fast. Rate of change of internal thermal energy, $\left(C_p\rho V\right)_{al}\dfrac{dT_{al}}{d\theta}$, in the aluminum container starts at 156 Watts at θ = 3 s, goes through a maximum of 383 Watts at $\theta = 14\,s\ or\ 0.23\,min$,

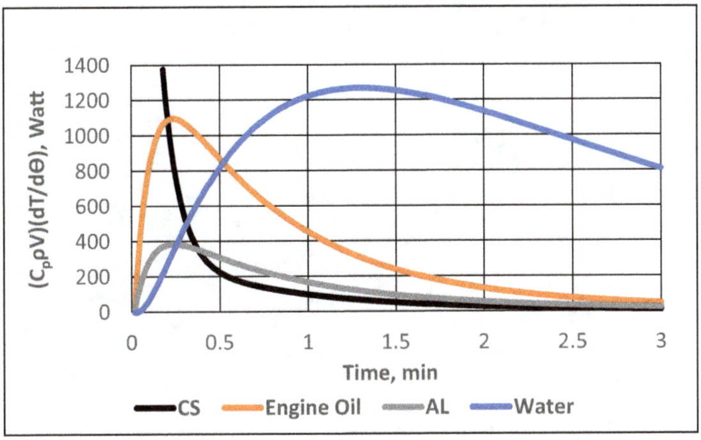

Figure 3-6 Rate of change of internal thermal energy for each section or lump

and also decreases fast. Rate of change of internal thermal energy, $\left(C_p\rho V\right)_{wl}\dfrac{dT_w}{d\theta}$, in water start at 9 Watts at $\theta = 3\ s$, goes through a maximum of 1,267 Watts later at $\theta\ = 78\,s\ or\ 1.3\,min$, and decreases slowly because of water's high capacitance.

4

Use of Ablation Materials As Heat Shield to Protect Spacecrafts Entering Earth's Atmosphere From Incoming Excessive Heat Loads

The surface boundary of a spacecraft entering Earth's atmosphere from space at very high speeds, i.e. about 18,000 mph, experiences high heat loads, i.e. in the order of 10^7 W/m^2, due to convection and radiation heat transfers and aerodynamic heating. Ablating materials are used to capture most of the incoming heat load to protect the spacecraft, its occupants and its instruments from excessive heat loads. Ablation of special materials such as teflon, phenolic nylon, and silicon based ablative materials are used to cover surfaces of a spacecraft in order to absorb, melt and evaporate during this high heat load event. These ablative materials have special characteristic such as low thermal conductivity, low density, low coefficient of thermal expansion, high emissivity and high melting temperatures.

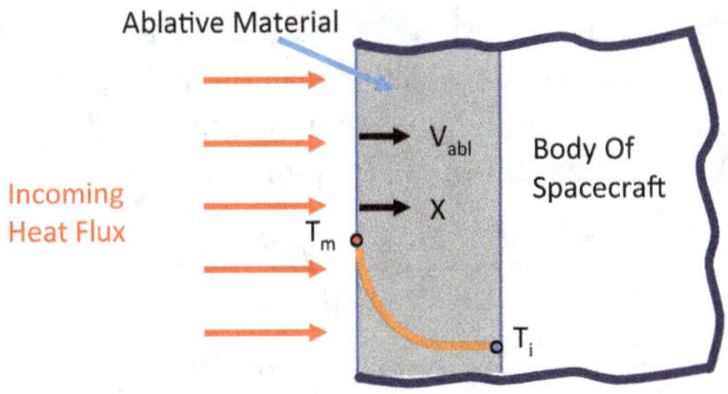

Figure 4-1 Body of a spacecraft covered with ablative material to prevent it from incoming high heat loads. Front surface of the ablation material has reached the melting temperature, T_m.

Let us use the one-dimensional transient heat conduction equation where $X = 0$ is the outer surface of the ablative material before ablative material's outer surface starts to melt as shown in Figure 4-1. Un-melted surface of the ablative material moves with a constant velocity V_{abl} after ablative material's surface starts to melt.

Following assumptions are made so that the partial differential equation for one-dimensional transient heat conduction in a semi-infinite body can be used to simulate the ablation process after the front surface of the ablation material has reached the melting temperature, T_m.

1. After the melting temperature, T_m, is reached at $X = 0$, The heat shield's ablating surface front moves at a constant rate, V_{abl}.
2. As $X \rightarrow \infty$, the temperature of the semi-infinite body approaches its initial temperature T_i.
3. Thickness of the spacecraft material is large as compared to the depth of the ablative material during the ablation melting process.
4. The spacecraft maintains a constant speed during re-entry.
5. Total energy that is absorbed by the ablative material during the re-entry stays a constant.
6. Melting and mass transfer processes in the ablation layer are neglected.
7. The temperature profile in the ablative material is obtained by using a moving coordinate system at a constant rate, $W = X - V_{abl}\theta$, where X is from the front face of the ablative material at time $\theta = 0$. $V_{abl}\theta$ is the length of the ablative material removed up to time θ.
8. Since the ablative material have low thermal conductivity, ablation occurs at the moving surface of the ablative material and T_m at the surface drops fast to T_i in thickness of the leftover ablation material as shown in Figure 4-1.
9. Assume constant ablative material properties during the ablation process.

The partial differential equation for one-dimensional transient heat conduction in a semi-infinite body is given in Equation 4-1, see Ref. [14].

$$\frac{\partial^2 T}{\partial X^2} = \frac{1}{\alpha}\frac{\partial T}{\partial \theta} \qquad \text{4-1}$$

where T in C is the temperature of the semi-infinite body at location X in m and at time θ in s. α is the thermal diffusivity of the ablative material in m²/s.

The temperature profile in the ablative material in Equation 4-1 is reduced to an ordinary differential equation by using a moving coordinate system at a constant rate, $W = X - V_{abl}\theta$, as shown in Equation 4-2.

$$\frac{d^2 T}{dW^2} = -\frac{V_{abl}}{\alpha}\frac{dT}{dW} \qquad \text{4-2}$$

Equation 4-2 has the following boundary conditions given in Equation 4-3.

$$T = T_m \text{ at } W = 0 \qquad \text{and} \qquad T = T_i \text{ as } W \rightarrow \infty \qquad \text{4-3}$$

The ordinary differential equation given in Equation 4-2 has the following general solution shown in Equation 4-4.

$$T = C_1 + C_2\, exp\left(-\frac{V_{abl}W}{a}\right) \tag{4-4}$$

where $C_1 = T_i$ and $C_2 = T_m - T_i$. Then, the temperature distribution in the semi-infinite slab takes the form shown in Equation 4-5.

$$\frac{T - T_i}{T_m - T_i} = exp\left(-\frac{V_{abl}W}{a}\right) \tag{4-5}$$

The ablative material's melting surface moves with a constant velocity V_{abl} as shown in Figure 4-2.

Let us know look at to the energy balance at the moving ablation surface. The incoming high heat load to the moving ablation surface either conducts through the ablative material or melts the ablative material as shown in Equations 4-6.

$$\left(\frac{q}{A}\right)_{in-total} = \left(\frac{q}{A}\right)_{conduction} + \left(\frac{q}{A}\right)_{ablation} \tag{4-6a}$$

$$\left(\frac{q}{A}\right)_{in-total} = -k\left(\frac{dT}{dW}\right)_{W=0} + \rho V_{abl} H_{abl} \tag{4-6b}$$

$$\left(\frac{q}{A}\right)_{in-total} = \rho C_p V_{abl}\left(T_m - T_i\right) + \rho V_{abl} H_{abl} \tag{4-6c}$$

Constant ablation velocity can be obtained from Equation 4-6c and it is given in Equation 4-7.

$$V_{abl} = \frac{\left(\frac{q}{A}\right)_{in-total}}{\rho H_{abl}\left(1 + \frac{C_p\left(T_m - T_i\right)}{H_{abl}}\right)} \tag{4-7}$$

where C_p is the specific heat of the ablative material in (W-s)/(kg-K). ρ is the density of the ablative material in kg/m³. H_{abl} is the heat of ablation for the ablative material in (W-s)/kg.

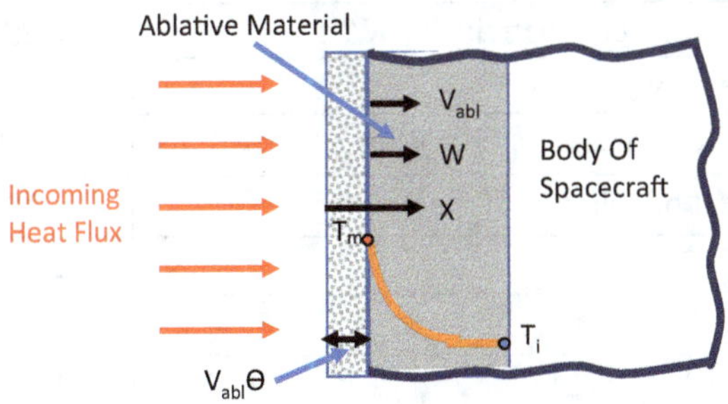

Figure 4-2 Body of a spacecraft covered with ablative material in a moving coordinate system at a constant rate, $W = X - V_{abl}\theta$.

Now let us define the ratio of the conduction energy that is going into the ablative material to the heat of ablation energy for ablative material as, E_{ratio}, which is given in Equation 4-8.

$$E_{ratio} = \frac{C_p\left(T_m - T_i\right)}{H_{abl}} \tag{4-8}$$

Using the definition of, E_{ratio}, we obtain ratios of three energies that govern the ablation process as shown in Equations 4-9.

$$\frac{\left(\frac{q}{A}\right)_{ablation}}{\left(\frac{q}{A}\right)_{in-total}} = \frac{1}{1+E_{ratio}} \tag{4.9a}$$

$$\frac{\left(\frac{q}{A}\right)_{conduction}}{\left(\frac{q}{A}\right)_{in-total}} = \frac{E_{ratio}}{1+E_{ratio}} \tag{4-9b}$$

$$\frac{\left(\frac{q}{A}\right)_{conduction}}{\left(\frac{q}{A}\right)_{ablation}} = E_{ratio} \tag{4-9c}$$

First, let us perform a temperature distribution case study using teflon as the ablation material. It is convenient to introduce a characteristic length, $L_{ch} = \frac{a}{V_{abl}}$, into Equation 4-5 as shown in Equation 4-10.

$$\frac{T - T_i}{T_m - T_i} = exp\left(-\frac{W}{L_{ch}}\right) \tag{4-10}$$

Constant teflon properties, the initial spacecraft temperature and incoming high total heat load that are used to determine the ablative material's temperature distribution are given in Table 4-1.

Thermal conductivity, k, W/(m-K)	0.22
Density, ρ, kg/m^{3k0}	1,922
Specific heat, C_p, (W-s)/(kg-K)	1,671
Heat of ablation, H_{abl}, (W-s)/kg	$7x10^6$
T_m, C	368
T_i, C	25
$(q/A)_{in-total}$, W/m^2	$7.24x10^6$

Table 4-1 Parameters that are used in the present analyses

Using the above parameters,

$$V_{abl} = 0.497\ mm/s; \quad E_{ratio} = 0.0819; \quad L_{ch} = 0.138\ mm$$

Temperature distribution in the ablative teflon material is determined using Equation 4-10 and it is shown in Figure 4-3 in orange color where L_{ch}=0.138 *mm*. If the ablation velocity of the ablative material can be doubled without changing its specific heat value, then L_{ch}= 0.069 *mm* and this temperature distribution is represented by the red line in Figure 4-3. L_{ch} is halved and therefore the thermal conductivity of this ideal ablative material also has to be halved to 0.11 W/(m-K). If the ablation velocity of the ablative material can be halved without changing its specific heat value, then L_{ch}= 0.276 *mm* and this temperature distribution is represented by the blue line in Figure 4-3. L_{ch} is doubled and therefore the thermal conductivity of this ideal ablative material also has to be doubled to 0.44 W/(m-K).

In temperature distributions in Figure 4-3, when $W = L_{ch}$ in the ablative material, non-dimensional temperatures drop from 1.0 to 0.367 and T goes down from 368 C to 150.9 C.

$\dfrac{\left(\frac{q}{A}\right)_{ablation}}{\left(\frac{q}{A}\right)_{in-total}}$ in Equation 4-9a and $\dfrac{\left(\frac{q}{A}\right)_{conduction}}{\left(\frac{q}{A}\right)_{in-total}}$ in Equation 4-9b are shown

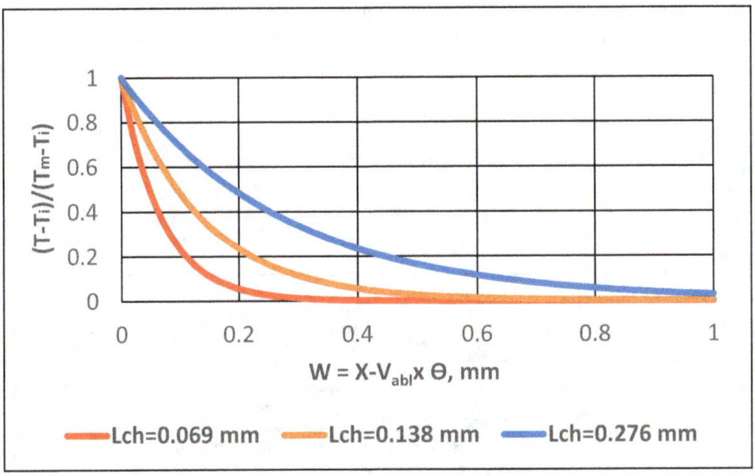

Figure 4-3 Temperature distribution in teflon ablative material (orange colored curve) compared with the temperature distributions with halving (red colored curve) and with doubling (blue colored curve) thermal conductivities of Teflon

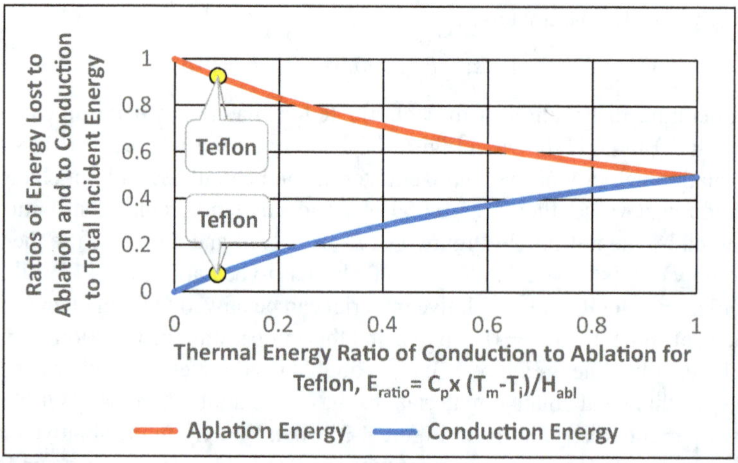

Figure 4-4 Ratios of energy lost to ablation and to conduction to total incident energy versus E_{ratio}. Ablative material teflon's location is shown in yellow dots.

as a function of E_{ratio}, which is the ratio of conduction energy to heat of ablation energy, in Figure 4-4. When all of the incoming energy goes into ablation energy, there is no conduction energy. When E_{ratio} is unity, half of the incoming energy goes to ablation energy and the other half to conduction energy. For the teflon material, 90.3% of the incoming energy goes to ablation energy and 9.7% of the incoming energy goes to conduction energy. The object function of the ablative material is to minimize the conduction energy.

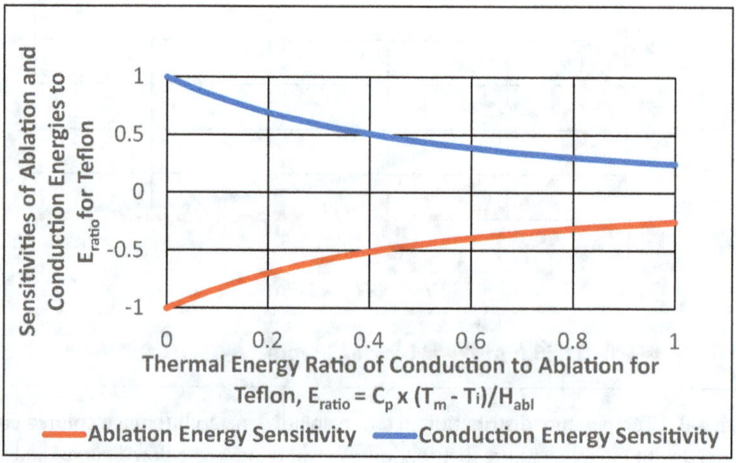

Figure 4-5 Sensitivities of ablation and conduction energies to E_{ratio} versus E_{ratio}

Sensitivities of ablation and conduction energies to E_{ratio} versus E_{ratio} are shown in Figure 4-5. Absolute values of sensitivities of both ablation and conduction energies are equally high to changes in low values of E_{ratio}.

Absolute values of sensitivities of both ablation and conduction energies equally asymptote to 0.25 as $E_{ratio} \to 1.0$.

Design of ablating materials for a spacecraft should be targeted to capture more than 90% of the incoming high heat load to protect the spacecraft, its occupants and its instruments from excessive incoming total heat loads, namely $E_{ratio} < 0.11$.

5

Stress Relieving a Copper Alloy Rod by Radiant Heating When Temperature Gradients Within The Rod are Negligible

In an electrical motors factory, after machining, copper alloy rods are stress relieved at a temperature of 260 C for 1 hour. Stress relieving is done in a vacuum oven to avoid surface oxidation and the vacuum oven's inner surfaces have radiant heating elements. A copper alloy cylindrical rod which has a characteristic length, $L_{cu} = \frac{V_{cu}}{A_{cu}} = 0.5R_{cu}$, or 0.005 m, has to be stress relieved by placing it in the middle of a vacuum oven with radiation heaters at its inner surfaces as shown in Figure 5-1.

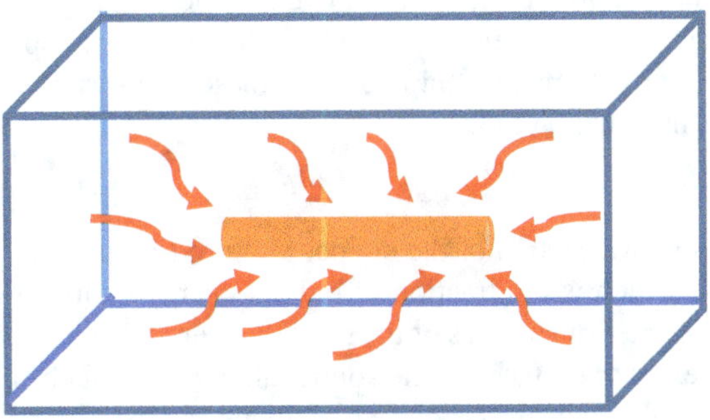

Figure 5-1 Stress relieving of a copper alloy cylindrical rod which is placed in the middle of a vacuum oven with radiation heaters at its inner surfaces

We will assume that the temperature within the copper alloy rod is uniform during the transient radiant heating process, namely there are no or very little temperature gradients within the copper alloy rod. The rate of internal energy increase in the copper alloy rod is from the radiant energy gain from its surfaces. Resistance to conductance heat transfer within the copper rod is much less than to the resistance of radiant heat transfer across surfaces of the copper alloy rod as given in Equation 5-1.

$$\frac{L_{cu}}{k_{cu}} << \frac{1}{h_r} \text{ or Biot \#} = \frac{h_r}{k_{cu} / L_{cu}} << 1 \qquad \text{5-1}$$

$L_{cu} = \frac{V_{cu}}{A_{cu}}$ is the characteristic length of the copper alloy rod in meters which is the ratio of its volume to its surface area. k_{cu} is the thermal conductivity of the copper alloy rod in W/(m-K). The radiation heat transfer coefficient, h_r, is defined as shown in Equation 5-2 and its dimension is W/(m^2-K).

$$h_r = \varepsilon_{cu}\, \sigma \left(T_{oven}^4 - T_{cu}^4\right) / \left(T_{oven} - T_{cu}\right) \qquad \text{5-2}$$

Oven's inner surfaces are isothermal surfaces irradiating as a blackbody, namely as σT_{oven}^4. σ is the Stefan-Boltzmann constant at 5.67x10^{-8} W/(m^2-K^4) and T_{oven} is the oven surface temperature in K. The copper alloy rod's surfaces are assumed to be gray surfaces where radiation emissivity, ε_{cu}, is equal to the radiation absorptivity, α_{cu}, namely $\varepsilon_{cu} = \alpha_{cu}$. Infrared radiation heat transfer from inner surfaces of the vacuum oven which completely surround exterior surfaces of the much smaller copper alloy rod is given in Equation 5-3.

$$Q_{oven \rightarrow cu} = A_{oven} F_{oven \rightarrow cu}\, \sigma \left(T_{oven}^4 - T_{cu}^4\right) \qquad \text{5-3}$$

A_{oven} is the interior surface area of the vacuum oven in m^2. $F_{oven-cu}$ is the configuration factor which is the fraction of radiant energy emitted diffusively and incident from the interior surfaces of the oven to the copper alloy rod which is located at the center of the oven. T_{cu} is the temperature of the copper alloy rod in K. There is a reciprocity relationship between these two surfaces' configuration factors as given in Equation 5-4.

$$A_{oven} F_{oven \rightarrow cu} = A_{cu} F_{cu \rightarrow oven} \qquad \text{5-4}$$

A_{cu} is the exterior surface area of the copper alloy rod in m^2. $F_{cu-oven}$ is the configuration factor which is the fraction of radiant energy emitted diffusively and incident from the exterior surfaces of the small copper alloy rod to oven's interior surfaces. Since exterior surfaces of the copper alloy rod is much smaller than the interior surfaces of the oven, $F_{cu \rightarrow oven} = 1$. Radiant energy coming from interior surfaces of the oven which is absorbed by copper alloy rod's surfaces is obtained by combining Equations 5-2 and 5-3 and it is given in Equation 5-5.

$$Q_{oven\rightarrow cu} = A_{cu}\varepsilon_{cu}\sigma\left(T_{oven}^4 - T_{cu}^4\right) \quad 5\text{-}5$$

Transient state of the copper alloy rod's temperature is the change in its internal energy by time, θ, in seconds, being equal to radiation energy received at its surfaces as shown in Equation 5-6.

$$\rho_{cu}C_{p-cu}V_{cu}\frac{dT_{cu}}{d\theta} = A_{cu}\varepsilon_{cu}\sigma\left(T_{oven}^4 - T_{cu}^4\right) \quad 5\text{-}6$$

The initial condition to solve the differential Equation 5-6 is given in Equation 5-7.

$$T_{cu} = 20C \ or \ 293K \ at \ \theta = 0 \quad 5\text{-}7$$

Also, copper alloy rod's properties in Equations 5-1 and 5-6, k_{cu}, ρ_{cu}, C_{p-cu}, ε_{cu}, L_{cu}, are assumed to be constants in time and they are valued at the average temperature of $T_{cu-ave} = \frac{(20+260)}{2} = 140C \ or \ 413K$. After separating variables T_{cu} and θ in Equation 5-6 takes the following integral form in Equation 5-8. $C1$ is the constant of integration in Equation 5-8.

$$\int\frac{dT_{cu}}{T_{oven}^4 - T_{cu}^4} - \frac{\varepsilon_{cu}\sigma}{\rho_{cu}C_{p-cu}L_{cu}}\int d\theta + C1 = 0 \quad 5\text{-}8$$

Integration of Equation 5-8 results in the following Equation 5-9.

$$\left(\frac{1}{4T_{oven}^3}\right)\left[Ln\left|\frac{T_{oven}+T_{cu}}{T_{oven}-T_{cu}}\right| + 2arctan\left(\frac{T_{cu}}{T_{oven}}\right)\right] - \frac{\varepsilon_{cu}\sigma\theta}{\rho_{cu}C_{p-cu}L_{cu}} + C1 = 0 \quad 5\text{-}9$$

Using the initial condition that is given in Equation 5-7 and a constant temperature for interior surfaces of the oven, $T_{oven} = 400C \ or \ 673K$, and also, multiplying all three terms in Equation 5-7 by $4T_{oven}^3 = 4\times 673^3$ gives the following Equation 5-10.

$$\left[Ln\left|\frac{673+T_{cu}}{673-T_{cu}}\right| + 2arctan\left(\frac{T_{cu}}{673}\right)\right] - 3.253 - 4\times 673^3\left(\frac{\varepsilon_{cu}\sigma\theta}{\rho_{cu}C_{p-cu}L_{cu}}\right) = 0 \quad 5\text{-}10$$

Properties of the copper alloy rod are given as:

$$k_{cu} = 50\frac{W}{m-K}, \ \rho_{cu} = 8{,}800\frac{kg}{m^3}, \ C_{p-cu} = 400\frac{J}{kg-K},$$

$$\varepsilon_{cu} = 0.5 \ , \ L_{cu} = 0.005m.$$

Then, Equation 5-10 takes the form shown in Equation 5-11.

$$f\left(T_{cu},\theta\right)=\left[Ln\left|\frac{673+T_{cu}}{673-T_{cu}}\right|+2arctan\left(\frac{T_{cu}}{673}\right)\right]-3.253-0.001964\theta=0 \qquad 5\text{-}11$$

where $f\left(T_{cu},0\right)=0$. Equation 5-11 is a transcendental equation whose roots for each value of time, θ, can be found when $f\left(T_{cu},\theta\right)=0$. Sample results for the present copper alloy rod's temperature up to the required 533 K or 260 C are shown in the following Table 5-1 and the copper alloy rod's temperature versus time profile is shown in Figure 5-2 for $\varepsilon_{cu}=0.5$.

θ, s	T_{cu}, K	T_{cu}, C
0	293	20
10	309	36
60	369	96
120	420	147
180	458	185
240	488	215
300	513	240
357.5	533	260

Table 5-1 Sample roots of $f(T_{cu},\theta) = 0$ obtained from Equation 5-11

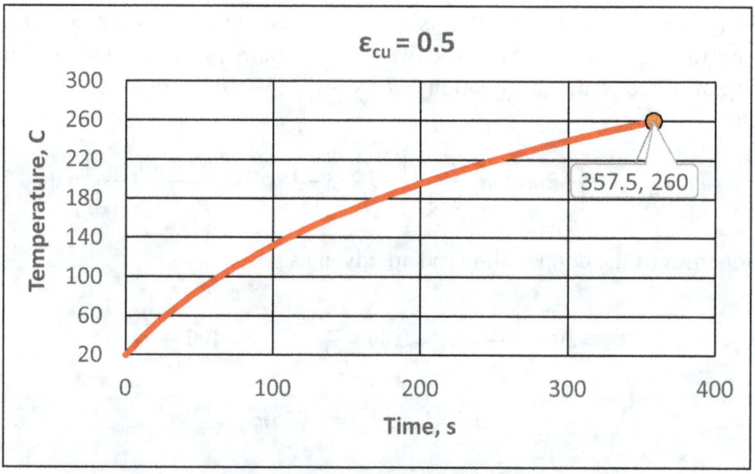

Figure 5-2 Copper alloy rod's temperature versus time profile until its temperature reaches the stress relieving temperature of 260 C in 357.5 seconds

We assumed in this analysis that the temperature within the copper alloy rod is uniform during the transient radiant heating process. Now, let us show that the resistance to conductance heat transfer within the copper alloy rod is much less than to the resistance of radiant heat transfer across its surfaces. During the radiant heating process, the radiation heat transfer coefficient reaches its maximum value at $T_{cu} = 533\,K\,or\,260\;C$. Using Equations 5-1 and 5-2,

$$h_r = \frac{0.5 \times 5.67 \times 10^{-8}\left(673^4 - 533^4\right)}{\left(673 - 533\right)} = 25.2 W/(m^2 - K)$$

$$Biot\ \# = \frac{h_r}{\frac{k_{cu}}{L_{cu}}} = \frac{25.2}{\frac{50}{0.005}} = 0.00252 \ll 1$$

The temperature within the copper rod is very close to being uniform during the transient radiant heating process as long as $Biot\ \# < 1$.

After the copper alloy rod's temperature reaches 260 C, the inner surface temperature of the vacuum oven is decreased to 260 C. The copper alloy rod stays in this environment for 1 hour and then the inner surface temperature heaters of the vacuum oven are turned off. The copper alloy rod cools down to room temperature in the vacuum oven before it is removed.

Copper alloy rod's emissivity and therefore absorptivity, ε_{cu}, changes between 0.4 to 0.6 due to its surface condition such as oxidation and its surface finish. Low emissivity values extend the time to reach 260 C. For example, when $\varepsilon_{cu} = 0.4$, it takes 429 seconds for the copper alloy rod to reach the stress relieving temperature. High emissivity values shorten the time to reach 260 C. For example, when $\varepsilon_{cu} = 0.6$, it takes 297.9 seconds for the copper alloy rod to reach the stress relieving temperature. These temperature versus time profiles with different emissivities are shown in Figure 5-3.

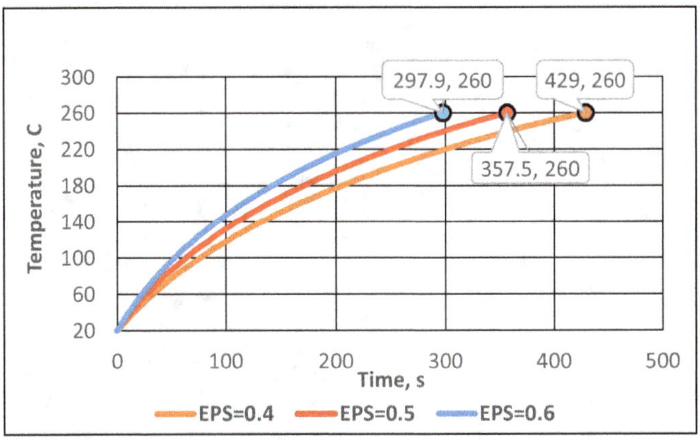

Figure 5-3 Copper alloy rod's temperature-time profiles for different emissivities

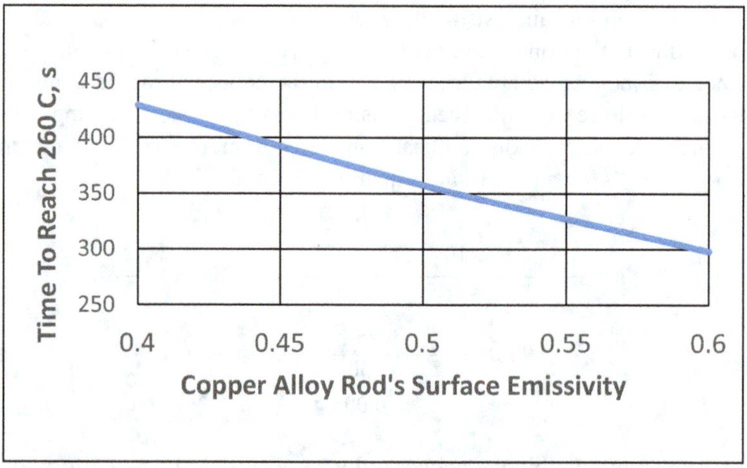

Figure 5-4 Time to reach 260 C versus copper alloy rod's emissivity

From temperature-time profiles presented in Figure 5-3, we can determine the sensitivity of change in time to reach 260 C to changes in copper alloy rod's emissivity which is the slope of the curve given in Figure 5-4. This sensitivity on the average is - 65.5 seconds decrease per 0.1 increase in emissivity.

6

Times to Hard Boil Different Sized Chicken Eggs

Chicken eggs have been studied extensively for their nutrients, physical and thermal properties. A large chicken egg is composed of a thin outside shell with a 0.03 cm mean thickness which is made out of calcium carbonate crystals and it is semipermeable. The middle layer is the egg white with a 1.22 cm mean thickness for a large egg that protects egg's yoke in the middle and the egg white is composed of water, protein, and ropes that hold the egg's yoke in the middle. A large egg's yoke in the middle with a 1.1 cm mean thickness contains water, proteins, vitamins and minerals. The egg shell has a high thermal conductivity of about 225 W/(m-K). When the egg is inserted into a boiling water at 100 C, the convection heat transfer coefficient is in the order of 10^3 W/(m^2-K) and therefore the Biot number for the egg shell is

$$Biot_{shell} = \frac{ht_{shell}}{k} \approx \frac{1,000 \times 0.0003}{225} = 0.0013$$

For egg's white and yoke, thermal conductivities are close to water's thermal conductivity at same temperature, about 0.54 and 0.34 W/(m-K), respectively.

Because of very low Biot number for the shell, temperature gradient through the shell during boiling can be neglected. Temperature gradients through the white and the yoke of different size chicken eggs will be analyzed assuming a spherically shaped egg which is initially at 5 C and which is suddenly immersed into a 100 C water surrounding it in order to hard boil it. Also, it is assumed that inside the egg, there are no temperature gradients in polar and azimuthal directions, namely all temperature gradients inside the egg are in the radial direction. For the present egg boiling analysis, exact solution to the transient and one-dimensional heat transfer equation, Equation 6-1 in spherical coordinates, will be used.

$$\frac{\partial^2 T}{\partial R^2} + \frac{2}{R}\frac{\partial T}{\partial R} = \frac{1}{\alpha}\frac{\partial T}{\partial \theta} \qquad \text{6-1}$$

where $T(R,\theta)$ is the temperature of the spherical egg in Kelvin or in centigrade. α is the thickness averaged, constant thermal diffusivity of egg's white and yoke in m^2/s with initial and boundary conditions given in Equations 6-2.

$$T_{ini}(R,0) = 5C \qquad \text{6-2a}$$

$$\left(\frac{\partial T}{\partial R}\right)_{R=0} = 0 \qquad \text{6-2b}$$

$$-k\left(\frac{\partial T}{\partial R}\right)_{R=R_o} = \bar{h}(T_w - T)_{R=R_0} \qquad \text{6-2c}$$

where k is the thickness averaged, constant thermal conductivity of egg's white and yoke in W/(m-K) and $\bar{h}$ is the average convection heat transfer at the surface of the egg in W/(m^2-K). For a sphere $T(R, \theta)$ that satisfy Equations 6-1 and 6-2, is given in Equations 6-3. See Ref. [10].

$$\frac{T(R,\theta) - T_{env}}{T_{ini} - T_{env}} = \sum_{i=1}^{\infty} A_i e^{-\lambda_i^2 FO} \left[\frac{\sin\left(\lambda_i \frac{R}{R0}\right)}{\left(\lambda_i \frac{R}{R0}\right)}\right] \qquad \text{6-3a}$$

$$A_i = \frac{4\left[sin(\lambda_i) - \lambda_i cos(\lambda_i)\right]}{2\lambda_i - sin(2\lambda_i)} \qquad \text{6-3b}$$

$$1 - \lambda_i cot(\lambda_i) = Biot\ Number \qquad \text{6-3c}$$

The center temperature for the egg's yoke is given in Equation 6-4 when $limit_{R\to 0} \frac{\sin\left(\lambda_i \frac{R}{R0}\right)}{\left(\lambda_i \frac{R}{R0}\right)} = \frac{0}{0} = 1$. The non-dimensional time, the Fourier number, for a sphere is $FO = \frac{a\theta}{R_0^2}$. The non-dimensional Biot number for a sphere, the ratio of conduction to convection resistances, is $\frac{\bar{h} R_o}{k}$.

$$\frac{T(R=0,\theta) - T_{env}}{T_{ini} - T_{env}} = \sum_{i=1}^{\infty} A_i e^{-\lambda_i^2 FO} \qquad \text{6-4}$$

According to egg boiling experiments, see Ref. [7], an egg's yoke becomes hard-boiled when its center temperature reaches 70 C. One and two-term approximations are used for the center temperature given by infinite series in Equations 6-3 and 6-4. For different size eggs from small to jumbo, both 1-term and 2-term approximations are applied to Equations 6-3 and 6-4. In all cases, 2-term approximation of center temperatures required 0.3% less time to reach hard boiled status than the 1-term approximation. Also, $FO \approx 0.2$ for all cases. So, center temperatures for different size eggs are obtained using the 1-term and 2-term approximations for Equations 6-3 and 6-4 which are shown in Equations 6-5.

$$\frac{T(R=0,\theta)-100}{5-100} = A_1 e^{-\lambda_1^2 FO} \quad \text{6-5a}$$

$$\frac{T(R=0,\theta)-100}{5-100} = A_1 e^{-\lambda_1^2 FO} + A_2 e^{-\lambda_2^2 FO} \quad \text{6-5b}$$

In the present study, chicken egg sizes are classified according to their weight and volume at an average constant density of 1.031 g/m³ as shown in Table 6-1.

In the following temperature calculations, egg's white thickness and egg's yoke thickness averaged thermal conductivity, k, and thermal diffusivity, α, are used as given below.

$$k = 0.43 \ W/(m-K) \qquad a = 1.32 \times 10^{-7} \ m^2/s$$

Surface averaged free convection heat transfer coefficient, $\bar{h}$, between egg's spherical surface immersed in 100 C water is calculated from the following empirical relationship given in Equation 6-6. See Ref. [10].

$$\overline{Nu_D} = 2 + \frac{0.589 Ra_D^{0.25}}{\left[1+(0.469/Pr)^{0.563}\right]^{0.444}} \quad \text{6-6}$$

where $\overline{Nu_D} = \frac{\bar{h}D}{k}$ is the Nusselt number or the surface averaged dimensionless temperature gradient at egg's spherical surface and $D = 2R_o$ is the diameter of the spherical egg in meters. $\bar{h}$ has the dimension W/(m²-K). $Pr = \frac{\nu}{a}$ is the Prandtl number which is the ratio of the momentum and thermal diffusivities.

EGG SIZE	AVERAGE WEIGHT, g	AVERAGE VOLUME, cm³	AVERAGE SPHERICAL RADIUS, R_o, cm
Small	42	40.74	2.135
Medium	49	47.53	2.247
Large	56	54.32	2.349
Extra Large	64	62.08	2.456
Jumbo	70	67.90	2.531

Table 6-1 Chicken egg size categories from small eggs to jumbo eggs

$Ra_D = \dfrac{g\beta\left(T_{Ro} - T_w\right)D^3}{\nu a}$ is the Rayleigh number which is the ratio of buoyancy forces to the product of momentum and thermal diffusivities. g is the gravitational constant at sea level, namely 9.81 m/s^2. β is the volumetric expansion coefficient in 1/K. All water properties are obtained from thermophysical property tables at a film temperature of $T_{w-film} = 0.5(373 + 278) = 325.5\,K$. Spherical surface averaged free convection heat transfer coefficients for different size eggs are shown in Figure 6-1.

From Figure 6-1, we can conclude that as the size of the egg increases so does the separation of boundary layer that is formed by water over the surface of the egg. Therefore, local free convection heat transfer coefficients decrease at separation of boundary layer and so does the surface averaged free convection heat transfer coefficients.

Temperatures distribution for a large egg when the center temperature is at 70 C is obtained using 1-term and 2-term approximations from Equations 6-3 and 6-4. Results are shown in Figure 6-2. 1-term and 2-term approximations are very close to each other, since $FO = 0.198$.

Since, an egg becomes hard-boiled when its yoke's center temperature reaches 70 C, we can determine the time it takes to hard boil a different size egg using 1-term approximation to Equations 6-3 and 6-4 which is shown in Equation 6-7.

$$\theta_{hard-boil} = -\left(\frac{R_o^2}{a\lambda_1^2}\right) Ln\left[\frac{\left(\dfrac{70-100}{5-100}\right)}{A_1}\right] \qquad \text{6-7}$$

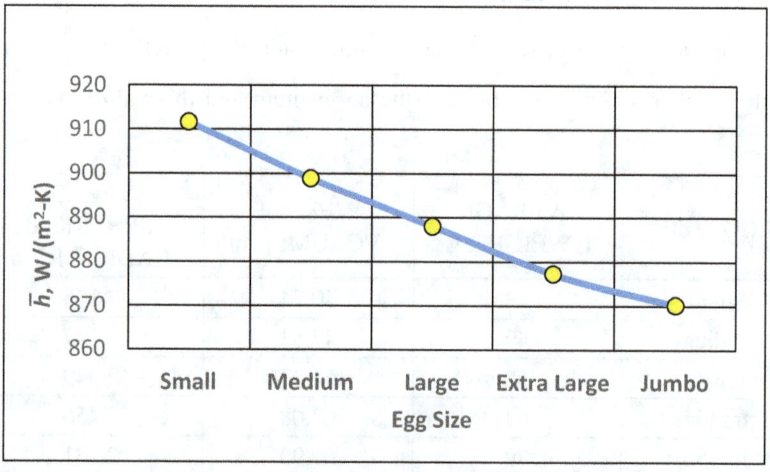

Figure 6-1 Spherical surface averaged free convection heat transfer coefficients for different size eggs

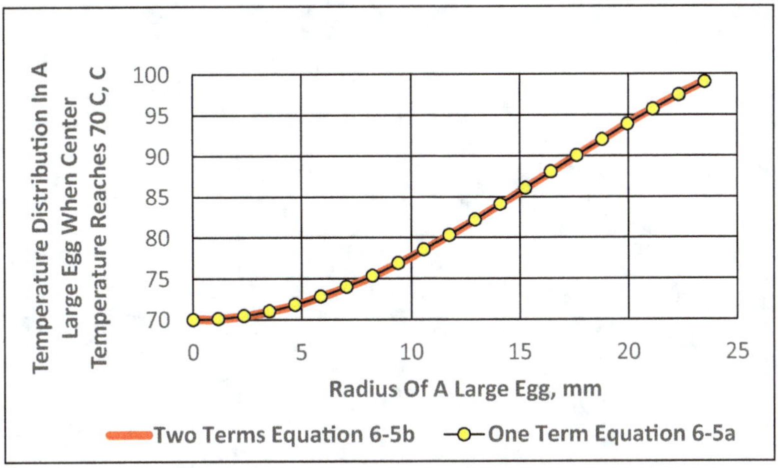

Figure 6-2 Temperatures distribution for a large egg when the center temperature is 70 C. Solid red line is the two-term approximation to exact Equations 6-3. Yellow dots are one-term approximation to exact Equations 6-3.

Figure 6-3 shows times to hard boil different size eggs from small ones in 11.4 minutes to jumbo ones in 16.0 minutes. At times shown in Figure 6-3, an egg should have a solid yoke and a solid white. Hard boiling times, under presently described conditions, change only 4.6 minutes from a small egg to a jumbo one.

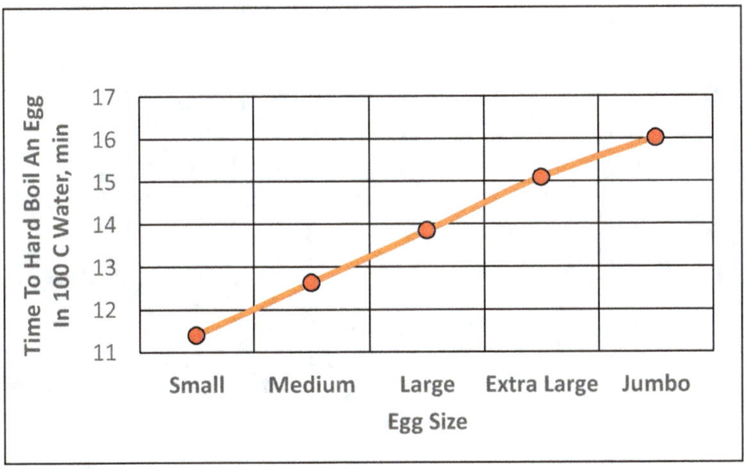

Figure 6-3 Times to hard boil different size eggs

7

Cooling a Pot of Soup in Air

In this chapter, the temperature-time history of a hot pot of soup which is immersed in still-air environment will be analyzed as shown in Figure 7-1. The still-air environment's temperature is controlled and it varies linearly with time. The hot soup in the thin-walled aluminum pot will have negligible internal conduction heat transfer resistance as compared to the external free convection heat transfer resistance to the still-air environment, namely the body of the hot soup will almost have a constant temperature during the cooling process. Free convection heat transfer coefficient between the still-air environment is assumed to be a constant during the whole cooling process and all air properties are calculated at film temperature.

The change in internal energy of the hot soup with respect to time equals the heat transfer from the hot soup to the still-air environment as shown in Equation 7-1.

$$-\rho_{soup} C_{p-soup} V_{soup} \frac{dT}{d\theta} = hA_{soup} \left[T - T_{air}(\theta) \right] \qquad 7\text{-}1$$

where the time dependent linear temperature variation of the still-air environment is given by Equation 7-2.

$$T_{air}(\theta) = A + B\theta \qquad 7\text{-}2$$

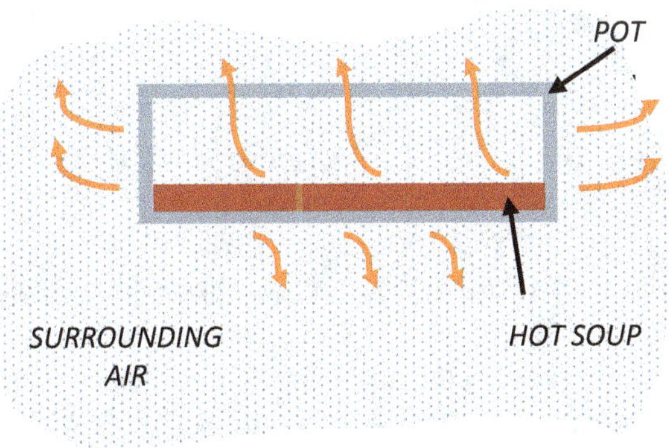

Figure 7-1 A hot pot of soup which is immersed in a still-air environment in which the air temperature varies with time

The energy balance in Equation 7-1 assumes a uniform temperature in the soup-pot combination at any time. For this assumption to be true, we have to check the ratio of conduction resistances in the soup-pot body with respect to convection resistance at soup-pot surfaces, namely the Biot number,

$Bi = \frac{L_{soup} / k_{soup}}{(1/h)_{soup-pot\ surfaces}}$. In order to be able to neglect temperature gradients within soup-pot body, Equation 7-3 has to be satisfied.

$$Bi = \frac{L_{soup} / k_{soup}}{(1/h)_{soup-pot\ surfaces}} < 0.1 \qquad \text{7-3}$$

Also, in the present analysis, we assume that the aluminum pot's temperature is the same as the hot soup's temperature, namely there is no temperature gradient through the 0.5 mm thick aluminum pot. In addition, the thermal capacity of the aluminum pot with very thin walls is negligible as compared to the hot soup's thermal capacity. The initial temperature of the hot soup is 100 C. In the present analysis, we will determine the time that it will take the 2-liter hot soup's temperature to go down from 100 C to 50 C. So, the average temperature of the hot soup and the aluminum container during this cooling period is 75 C at which the following constant thermophysical properties are used. In order to verify this assumption, we have to compare the conduction heat transfer resistance of soup's geometry with a characteristic length, L_{soup}, to the conduction heat transfer resistance of the aluminum with a pot thickness, t_{al}.

Pot Diameter = 40 cm; *Pot Height = 5 cm;* *Pot Thickness = 0.5 mm*

Hot Soup Volume V_{soup} *= 2 liters;* $\frac{V_{soup}}{A_{soup}} = L_{soup} = 7.4\,mm$

$$k_{soup} = 0.668\ W/(m-K) \quad ; \quad k_{al} = 168\ W/(m-K)$$

$$R_{al} = \frac{t_{al}}{k_{al}} = 3.0 \times 10^{-6}\left(m^2-K\right)/W$$

$$R_{soup} = \frac{L_{soup}}{k_{soup}} = 1.1 \times 10^{-2}\left(m^2-K\right)/W$$

$\frac{R_{al}}{R_{soup}} = 0.027$ So, the assumption that was made above about the aluminum pot's temperature being the same as the hot soup's temperature is valid. The assumption is about neglecting the thermal capacity of the aluminum pot's walls is discussed next.

$\rho_{al} = 2,770\,kg/m^3$; $C_{p-al} = 875\,J/(kg-K)$;
$V_{al} = 9.42\times10^{-5}\,m^3$;
$Thermal\,Capacity_{al} = \rho_{al}C_{p-al}V_{al} = 228.4\,J/K$
$\rho_{soup} = 2,770\,kg/m^3$; $C_{p-soup} = 875\,J/(kg-K)$;
$V_{soup} = 0.002\,m^3$;
$Thermal\,Capacity_{soup} = \rho_{soup}C_{p-soup}V_{soup} = 8169.4\,J/K$

Thermal capacity of the aluminum pot is only 2.8% of thermal capacity of the hot soup. So, the thermal capacity of the aluminum pot can be neglected in the present analysis.

The free convection heat transfer coefficient, h, used for soup-pot surfaces in Equation 7-1 is obtained from the empirical Equation 7-4, see Ref. [10], the Nusselt number.

$$Nu_{Lsoup} = \frac{hL_{soup}}{k_{air}} = 0.27\,Ra_{Lsoup}^{0.25} \qquad 7\text{-}4$$

where $Ra = \dfrac{g\beta\left(T-T_{air}\right)_{average}L_{soup}^3}{\nu a}$ is the Rayleigh number. In present calculations, the still air temperature, Equation 7-2, is at 100 C at $\theta = 0\,s$ and decreases with a slope of –0.1, namely $T_{air}(\theta) = 100 - 0.1\theta$. So, air's thermophysical properties are used at a film temperature of 62.5 C or 335.5 K. The resulting value of free convection heat transfer coefficient is, $h = 9.5\,W/\left(m^2-K\right)$.

We can now calculate the Biot number given in Equation 7-3.

$$Bi = \frac{L_{soup}/k_{soup}}{(1/h)_{soup-pot\ surfaces}} = \frac{9.5\times0.0074}{0.668} = 0.105$$

We can now use Equation 7-1 after assuming uniform temperature in the soup-pot combination at any time between $50C \le T \le 100C$. In Equation 7-4, $h \propto L_{soup}^{-0.25}$. Then, for lower volumes of hot soup as L_{soup} goes down, $Bi < 0.1$, since $Bi \propto L_{soup}^{0.75}$ as shown in Figure 7-2.

Now, let us solve the first order linear differential Equation 7-1 using the initial condition for the hot soup given in Equation 7-5.

$$T(0) = 100C \qquad 7\text{-}5$$

First step is to rewrite Equation 7-1 by introducing the time constant, τ, in seconds in Equation 7-6.

$$\tau = \frac{\rho_{soup}C_{p-soup}V_{soup}}{hA_{soup}} = \frac{\rho_{soup}C_{p-soup}L_{soup}}{h} \qquad 7\text{-}6$$

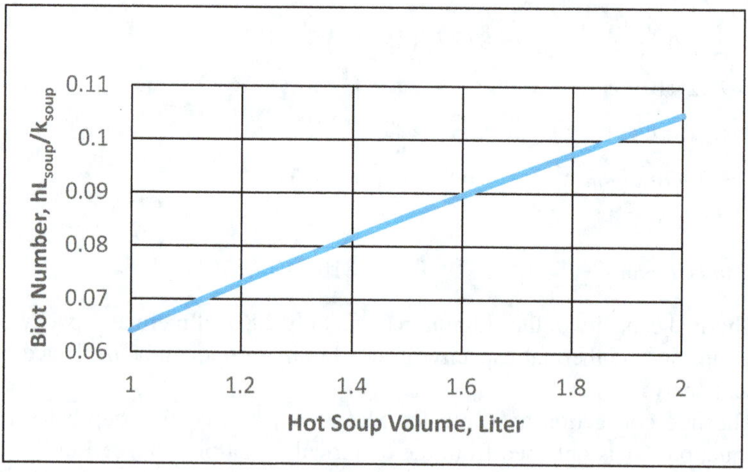

Figure 7-2 Biot number versus hot soup volume for cooling it in still-air environment's temperature varying linearly with time

Using Equations 7-2 and 7-6, Equation 7-1 takes the following differential form in Equation 7-7.

$$dT + \left(\frac{1}{\tau}\right) T d\theta - \left(\frac{1}{\tau}\right)(A + B\theta)\, d\theta = 0 \qquad \text{7-7}$$

Since the last term in Equation 7-7 is exact by itself, let us introduce an integration factor, $V(\theta)$, in order to make first two terms exact in Equation 7-8.

$$\frac{dV}{d\theta} = \frac{V}{\tau} \quad or \quad V = e^{\frac{\theta}{\tau}} \qquad \text{7-8}$$

Equation 7-7 takes the following form with the substitution of the integration factor to all three terms as shown in Equation 7-9.

$$V\frac{dT}{d\theta} + \left(\frac{dV}{d\theta}\right) T - V\left[\left(\frac{1}{\tau}\right)(A + B\theta)\right] = 0 \qquad \text{7-9}$$

Integrating Equation 7-9 term by term gives the following Equations 7-10.

$$e^{\frac{\theta}{\tau}} T - e^{\frac{\theta}{\tau}} A - e^{\frac{\theta}{\tau}} B(\theta - \tau) = C1 \qquad \text{7-10a}$$

or

$$T - A - B(\theta - \tau) = C1 e^{\frac{\theta}{\tau}} \qquad \text{7-10b}$$

Using the initial condition in Equation 7-5, the constant of integration, $C1$, becomes as shown in Equation 7-11.

$$C1 = T(0) - A + B\tau \qquad 7\text{-}11$$

Then, Equation 7-10 becomes as shown in Equation 7-12 which is the solution to Equation 7-1, 7-2 and 7-5 in which θ are τ in seconds.

$$T(\theta) = A + B(\theta - \tau) + \left[T(0) - A + \tau B\right]e^{\frac{\theta}{\tau}} \qquad 7\text{-}12$$

Next, let us look at to the temperature-time variation of a pot of hot soup using Equation 7-12 for different linear variations of still air environment's temperature with time as presented in Table 7-1.

Soup temperature versus time curves for different still-air environment temperatures that are outlined above in Table 7-1 are shown in the following Figure 7-3 for a hot soup volume of 2 liters with a time constant of 3,170 seconds. The initial temperatures of the hot soup and the still-air environment for all cases shown in Figure 7-3 are 100 C.

Hot soup cools faster as the still-air environment temperature cools faster, namely absolute value of B increases. In the $A = 0\; and\; B = 0$ case, hot soup cools down following the exponential decay curve. For example, the hot soup cools down to 50 C in $-3{,}170 \ln \frac{50}{100} = 2{,}197.3\,s\; or\; 36.6\,min$. In this exponential decay case, hot soup temperature will reach 36.8 C when $\theta = \tau$.

Let us know analyze the hot soup cooling cases from the initial temperature of 100 C down to 50 C with following constraints that the chef wants. He wants to take out the hot soup pot from his still-air-cooling environment after the hot soup reaches 50 C and serve it to his customers. His still-air-cooling environment equipment provides the following linear cooling profile shown in Equation 7-13. The hot soup is always in a standard aluminum pot that was specified above.

$$T_{air}(\theta) = 100 - 0.1\theta \qquad 7\text{-}13$$

A, C	B, C/s
0	0
100	–0.01
100	–0.02
100	–0.04
100	–0.10

Table 7-1 $T_{air}(\theta) = A + B\,\theta$ Variations in still-air environment temperature

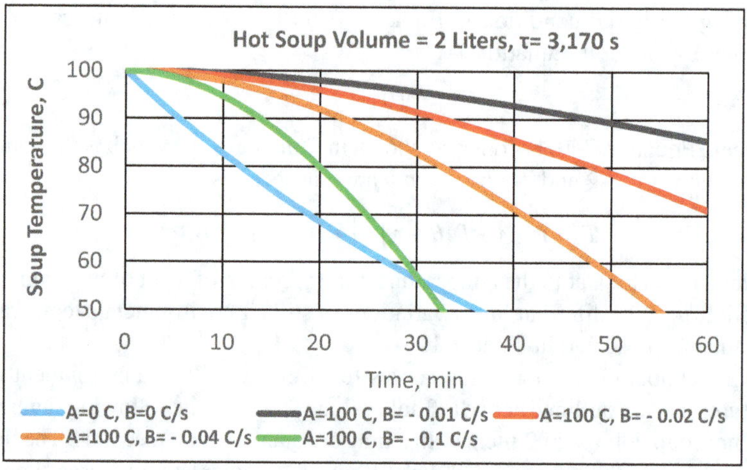

Figure 7-3 Soup temperature versus time curves for different still-air environment temperatures. The initial temperatures of the hot soup and the still-air environment for all cases are 100 C

So, the chef is going to use temperature-time history curves for different hot soup volumes given in Figure 7-4 to determine the hot soup's cooling times.

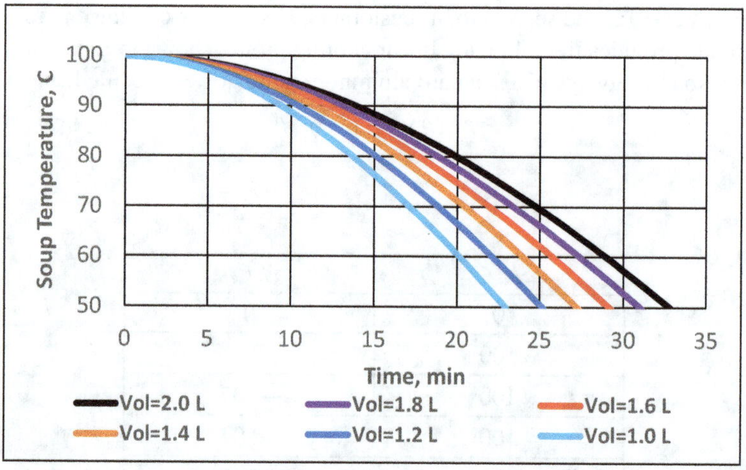

Figure 7-4 Hot soup's temperature-time histories from 100 C to 50 C in still-air linear cooling profile given in Equation 7-13 for various hot soup volumes

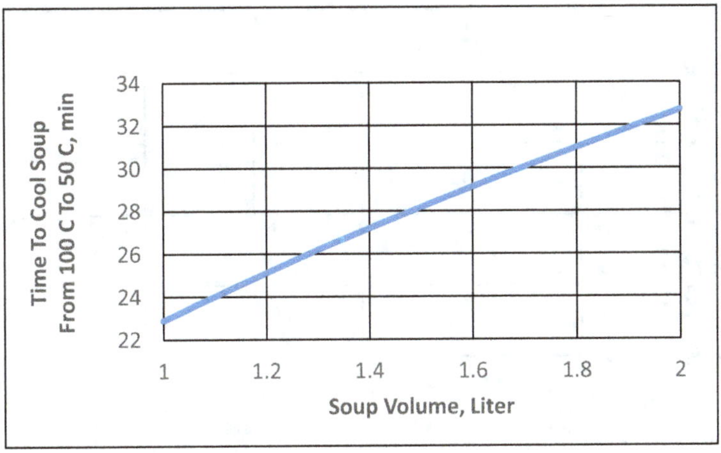

Figure 7-5 Hot soup cooling times from 100 C to 50 C for different soup volumes in still-air linear cooling profile given in Equation 7-13

Figure 7-5 might be easier for the chef which shows hot soup's cooling times from 100 C to 50 C for different soup volumes in still-air linear cooling profile given in Equation 7-13.

Data in Figure 7-5 is obtained by solving Equation 7-12 using different time constants, $\tau = \frac{\rho_{soup} C_{p-soup} L_{soup}}{h}$, related to different volumes of hot soup. Also, times that the hot soup reached 50 C were obtained by trial-and-error to temperature-time history curves at 50 C.

Next, let us look at to the sensitivity of time it takes to cool a pot of soup to its volume, namely $\frac{d\theta_{50}}{dV_{soup}}$, in still-air linear cooling profile given in Equation 7-13. As the hot soup's volume decreases, this sensitivity increases as shown in Figure 7-6. On the average $\frac{d\theta_{50}}{dV_{soup}}$ is 0.98 min/0.1 Liter of hot soup.

From the derivative of temperature with respect to time in Equation 7-12, we can determine the instantaneous rate of heat transfer from the hot soup as shown in Equation 7-14.

$$q = \rho_{soup} C_{p-soup} V_{soup} \frac{dT}{d\theta} \qquad \text{7-14}$$

$\frac{dT}{d\theta}$ in Equation 7-14 can be obtained from hot soup's temperature-time curves' slopes given in Figure 7-3. We can also determine $\frac{dT}{d\theta}$ from Equation 7-12 as shown in Equation 7-15.

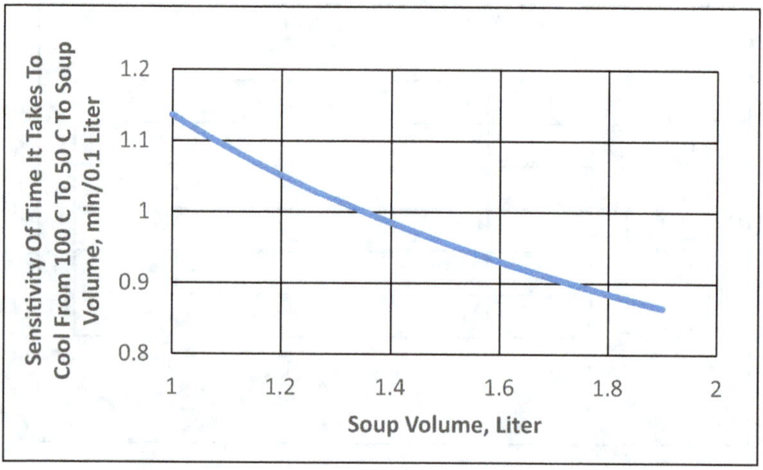

Figure 7-6 Sensitivity of time it takes to cool a pot of hot soup from 100 C to 50 C to its volume, $\frac{d\theta_{50}}{dV_{soup}}$, versus hot soup's volume

$$\frac{dT}{d\theta} = B - \left(\frac{1}{\tau}\right)\left[T(0) - A + \tau B\right]e^{\frac{\theta}{\tau}} \tag{7-15}$$

$\frac{dT}{d\theta}$ is shown for different hot soup volumes in Figure 7-7. Absolute values of hot soup's temperature-time curves' slopes increase as cooling times increase and as hot soup's volumes decrease.

Amount of internal energy lost from the hot soup in Joules when its temperature decreases from 100 C down to 50 C in time θ_{50} in a still-air cooling environment with the linear cooling profile shown in Equation 7-13 can be obtained by integrating the instantaneous rate of heat transfer given in Equation 7-14 as shown in Equation 7-16.

$$Q = \int_0^{\theta_{50}} q d\theta = \rho_{soup} C_{p-soup} V_{soup} \frac{dT}{d\theta} \tag{7-16}$$

Using Equation 7-12 and evaluating it between times 0 and θ_{50}, the amount of internal energy, Q, lost from the hot soup in Joules when its temperature decreases from 100 C down to 50 C, is shown in Equation 7-17.

$$Q = \rho_{soup} C_{p-soup} V_{soup} \left[B\theta_{50} + \left(T(0) - A + \tau B\right)\left(e^{\frac{\theta}{\tau}} - 1\right)\right] \tag{7-17}$$

Using the still-air-cooling environment with a linear cooling profile given in Equation 7-13 and Equation 7-17, the amount of internal energy lost from hot

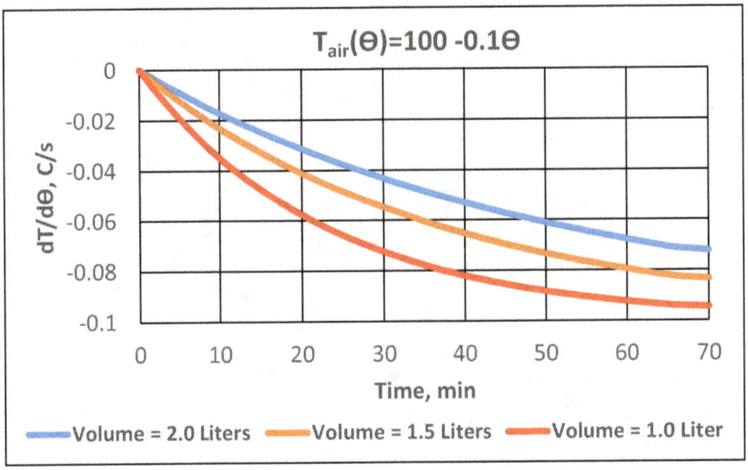

Figure 7-7 Derivatives hot soup's temperature with respect to time versus time for different hot soup volumes

soup in Joules when its temperature decreases from 100 C down to 50 C for different hot soup volumes is shown in Figure 7-8.

Internal energy, Q, lost from hot soup in Joules when its temperature decreases from 100 C down to 50 C for different hot soup volumes in still-air-cooling environment with a linear cooling profile shows almost a linear increase with increasing hot soup volume with an average slope of 20,434 J/0.1 Liter of hot soup.

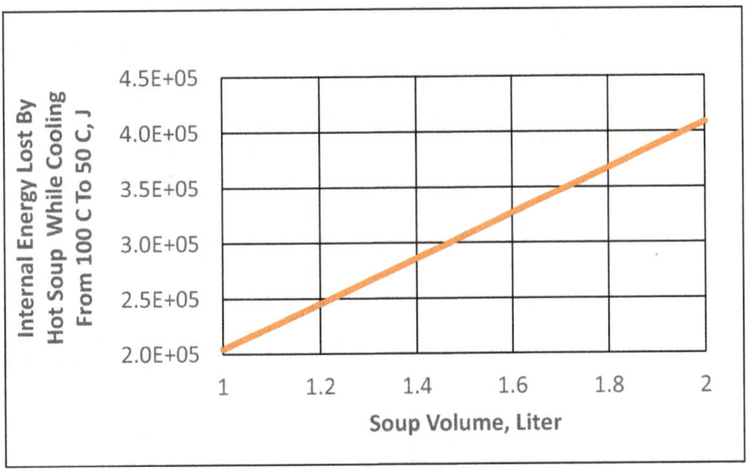

Figure 7-8 Internal energy, Q, lost from hot soup in Joules when its temperature decreases from 100 C down to 50 C for different hot soup volumes

8

Solar Energy Storage in Water

World energy consumption in 2040 is estimated to be 1.7×10^{17} W-hr/year. Solar energy will provide about 10% of this world energy consumption while replacing coal and oil. In this chapter, incoming solar energy that is processed in a solar collector to energize flowing water will be optimized for energy usage as shown in Figure 8-1. This solar energy usage optimization process starts with the knowledge of solar radiation energy that is available at a particular location and at a particular time on Earth. Hourly averaged direct and diffuse solar energy measured data, shown in Figure 8-2, that are incident on a horizontal surface at Fort Peck, Montana on 16 July 2015 will be used, see Ref. [12], (National Oceanic and Atmospheric Administration National Solar Radiation Database).

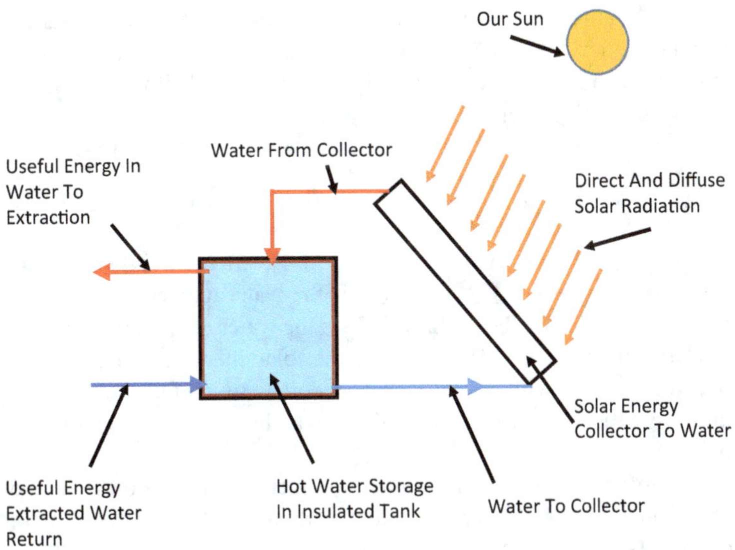

Figure 8-1 Solar energy storage system in water through a collector

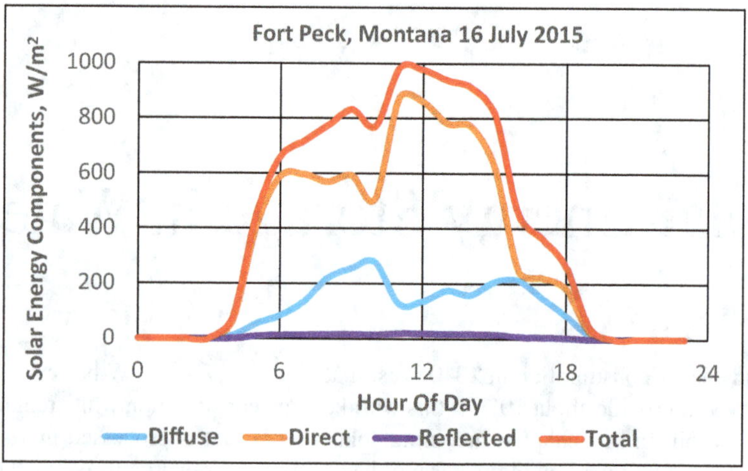

Figure 8-2 Location and date specific direct, diffuse, 2% reflected and total hourly averaged solar energies on a horizontal surface on a cloudless day at Fort Peck, Montana on 16 July 2015

In Figure 8-2, it was assumed that 2% of incoming direct and diffuse solar radiation was reflected from the collector's surface. Fort Peck, Montana is located at 48.3^0 North latitude. It is also assumed that water in the storage tank is well mixed and therefore there is no temperature stratification in the tank. Water flows in the system with forced circulation. The water tank and all water pipes are well insulated to minimize energy losses to the environment. Transient heat transfer energy balance for water in the storage tank is given in Equation 8-1.

$$\rho_w C_{p-w} V_w \frac{dT_w}{d\theta} = Q_{in-solar} - Q_{energy\ extraction} - Q_{out-env} \qquad 8\text{-}1$$

where $\rho_w V_w = 1,000\,kg$ is the water temperature averaged weight of water during the 24-hr day. $C_{p-w} = 4,181\ J/(kg - C)$ is the water temperature averaged specific heat of water during the 24-hr day. T_w is the water temperature in the storage tank in C. θ is time in seconds. $Q_{in-solar}$ is the solar energy transferred to water at the collector in Watts with a solar radiation collection surface area of 25 m^2 and an optimized collection surface tilt angle with an incoming solar energy multiplication factor of 1.11.

The incoming solar energy multiplication factor of 1.11 is obtained from sun rays elevation angle of 63.5^o and a collector tilt angle of 33° for Fort Peck, Montana on 16 July 2015.

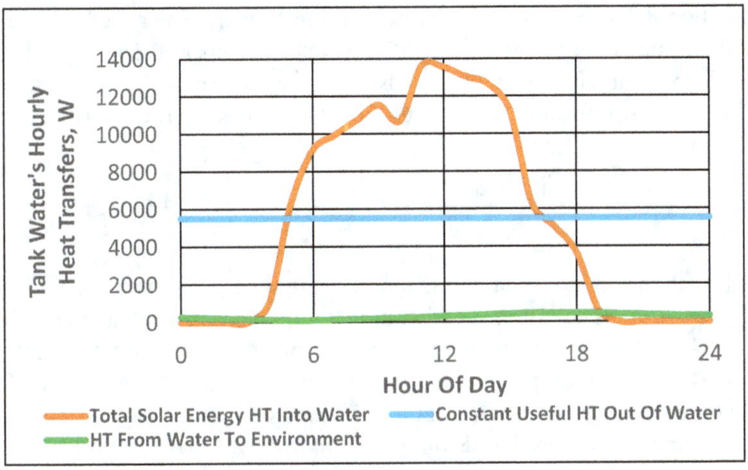

Figure 8-3 Tank water's hourly averaged heat transfers versus hour of day

Figure 8-3 shows the total hourly averaged solar energy that is transferred into water with a collector to water heat transfer efficiency of 50% in Fort Peck, Montana on 16th of July, 2015, namely $Q_{in-solar}$. Data shown in Figure 8-3 is the result of hourly averaged data given in Figure 8-2 with the inclusion of a 25 m^2 collector area factor, a 1.11 optimized collector tilt angle factor and a 0.5 collector to water heat transfer efficiency factor.

Figure 8-3 also shows the 24-hour constant extraction of useful heat from water, $Q_{energy\ extraction}$, in Watts such that at the beginning of each day, the water temperature in the tank is always at 60 C. About 5.1% of the energy in water is lost to the environment given by, $Q_{out-env}$. 24-hour summation of all energies involved in Figure 8-3 is zero, since a new day always starts at a water temperature of 60 C. $Q_{out-env}$ is obtained from series heat transfer resistances from tank water to the air environment as shown in Equation 8-2.

$$U = \frac{1}{\left(\frac{1}{h_{w-t}}\right)+\left(\frac{t_t}{k_t}\right)+\left(\frac{t_{ins}}{k_{ins}}\right)+\left(\frac{1}{h_{ins-air}}\right)} \qquad 8\text{-}2$$

Assuming free convection heat transfers between the water in the tank and the tank's inner surface and between the outer insulation layer surface of the tank and the environment air and assuming linear conduction heat transfers through tank's body material thickness and tank's outer insulation's thickness, the transmittance between water in the tank and the environment air comes out to be 7.0 W/C, namely $UA_{w-tank} = 7.0\ W/C$.

Equation 8-1 is solved numerically by using forward Euler method which is an explicit numerical method where $T_{w,j+1}$, water's temperature in the tank after a time increment $\Delta\theta = 3{,}600$ seconds. $T_{w,j+1}$ is determined from values that are already known from previous hour. Equation 8-1 is rewritten for the explicit numerical method in Equation 8-3.

$$T_{w,j+1} = T_{w,j} - \left(\frac{\Delta\theta}{\rho_w C_{p-w} V_w}\right)\left(Q_{in-solar} - Q_{energy\ extraction} + Q_{out-env}\right)_j \qquad 8\text{-}3$$

In Equation 8-3, constant extraction of useful energy from water during the day, $Q_{energy\ extraction}$, is determined by iteration to Equation 8-3 so that after 24 hours of extraction of useful energy, at the beginning of a day, the water temperature is always 60 C. The air temperature of the environment in these calculations is 20 C.

Water temperature in the tank at the beginning of an hour and at the end of an hour are shown in Figure 8-4. From midnight to 5 AM and from 4 PM to midnight, since there is none or little solar energy heat transfer into water, the water temperature at the end of the hour is always less than the water temperature at the beginning of the hour.

However, from 5 AM to 4 PM solar heat transfer into water exceeds the sum of heat transferred to energy extraction and to the environment. Therefore, the temperature of the water in the tank at the end of the hour is always more than the water temperature at the beginning of the hour. Water temperatures in the tank go through a minimum of 36.3 C at 4 AM and through a maximum of 87.6 C at 4PM as shown in Figure 8-4.

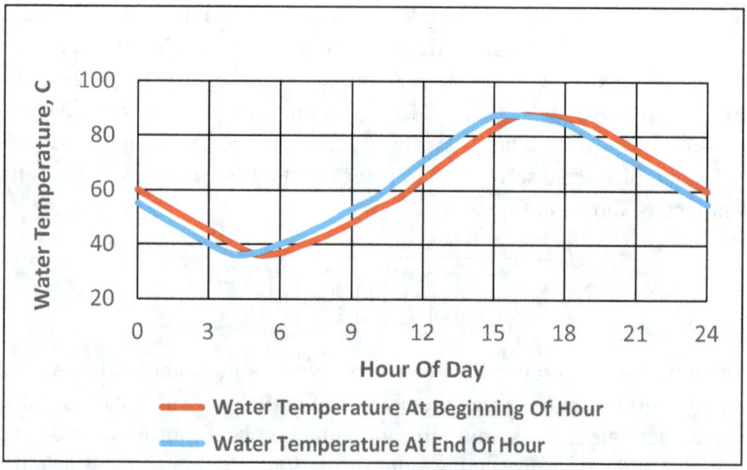

Figure 8-4 Water temperature variations in the tank during a day at beginning and end of an hour

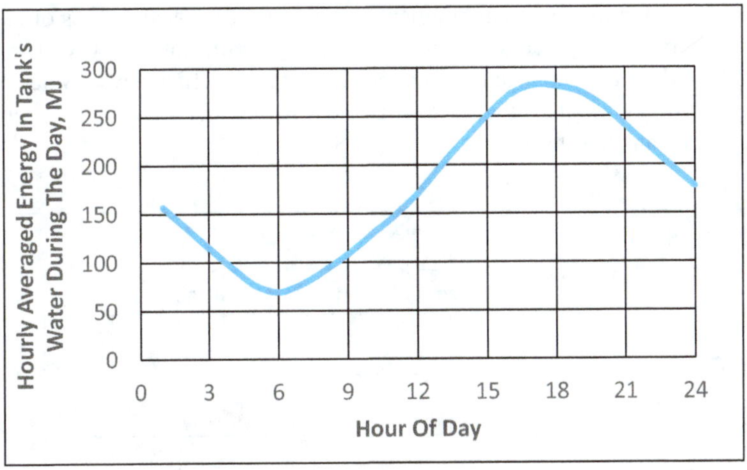

Figure 8-5 Hourly averaged energy in tank's water above 20 C environment temperature during the day

Hourly averaged energy in tank's water above 20 C is shown in Figure 8-5. This water energy, given in MJ above 20 C, is the average energy at the beginning and at the end of the hour. The minimum water energy is 69.4 MJ at 6 AM. The maximum water energy is 282.2 MJ at 5 PM.

Total solar energy going in to tank's water, out to energy extraction, and out to air environment per day versus collector's heat transfer efficiency to water is shown in Figure 8-6. Total solar energy going into water, then used for useful

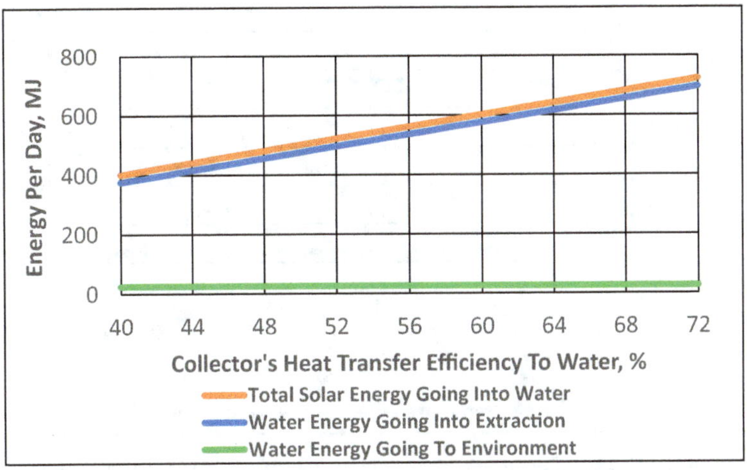

Figure 8-6 Total solar energy going into tank's water, out to energy extraction, and out to air environment per day versus collector's heat transfer efficiency to water

extraction and going to air environment are all almost linear functions of collector's heat transfer efficiency to water. Sensitivity of the solar energy change is about 100 MJ per 0.1 change in collector's heat transfer efficiency to water. Solar energy wasted going into the air environment is only 5.1% of the solar energy captured by tank's water.

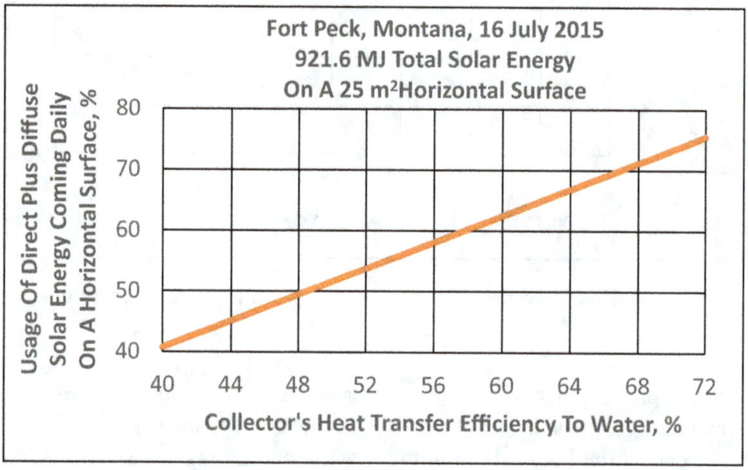

Figure 8-7 Percentage usage of direct plus diffuse daily total solar energy coming on a horizontal surface versus collector's heat transfer efficiency to tank's water

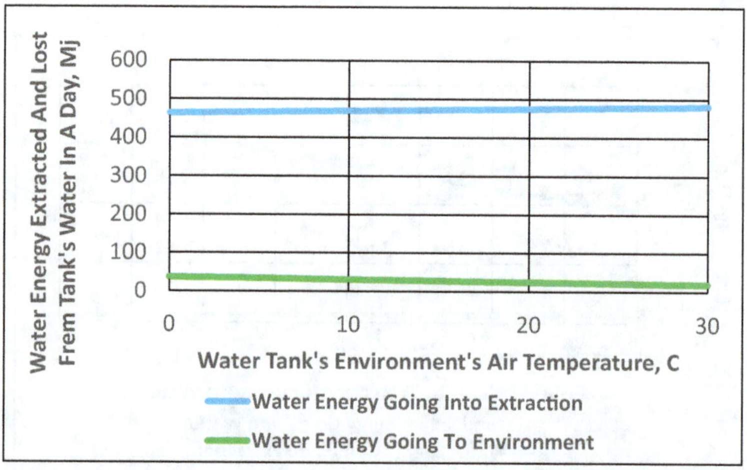

Figure 8-8 Effects of environment's air temperature to water's energy going into extraction and to environment in a day versus environment's air temperature

Percentage usage of direct plus diffuse daily total solar energy coming on a horizontal surface versus collector's heat transfer efficiency to tank's water is shown in Figure 8-7. Usage percentage of direct plus diffuse daily total solar energy coming on a horizontal surface and collector's heat transfer efficiency to tank's water are almost the same.

Effects of environment's air temperature to water's energy going into extraction and to environment's air as a function of environment's air temperature are shown in Figure 8-8. The sensitivity of change in extracted solar energy to change in environment's air temperature is + 0.6 MJ/C. The sensitivity of change in solar energy lost to environment to change in air temperature is - 0.6 MJ/C.

9

A Finite Difference Method to Solve 2-Dimensional Transient Heat Transfer Equation

The lumped capacitance method and analytical solutions cannot be applied to transient and multi-dimensional heat transfer problems with complicated geometries and boundary conditions. In this chapter, we will cover the explicit finite difference method application to a 2-dimensional transient heat transfer problem with multiple boundary conditions. Let us investigate the cooling of a stainless-steel rectangular prism bar with 1 m by 1 m square insulated ends from an initial temperature of 500 C down to 100 C at its center. The rectangular prism bar is subject to free and forced convection heat transfer at its four surfaces as shown in a 1 m by 1 m square cross-section view in Figure 9-1.

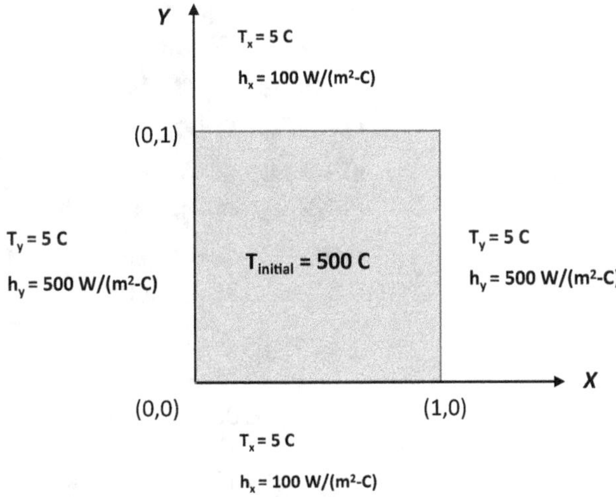

Figure 9-1 Cross-section of the stainless-steel rectangular prism bar

The 2-dimensional transient heat transfer equation with constant properties and with no internal energy generation is shown in Equation 9-1.

$$\frac{\partial^2 T}{\partial X^2} + \frac{\partial^2 T}{\partial Y^2} = \frac{1}{\alpha}\frac{\partial T}{\partial \theta} \qquad 9\text{-}1$$

where $T(X, Y)$ is the temperature of the stainless-steel rectangular prism bar at (X, Y) location in C. θ is time in s. α is stainless steel's thermal diffusivity in m^2/s. Explicit finite difference method is applied to Equation 9-1 by using conservation of energy at each (X, Y) node. Since there is symmetry in the heat transfer coefficients both in the X-direction and in the Y-direction, only a quarter of the (X, Y) surface from (0,0) to (0.5,0) in the X direction and (0,0) to (0, 0.5) in the Y direction shown in Figure 9-1 will be analyzed as illustrated in Figure 9-2.

In Figure 9-2, orange dash lines represent symmetry boundaries and nodes on these boundaries will be treated as insulated nodes, namely nodes (0,5), (1,5), (2,5), (3,5), (4,5), (5,5), (5,4), (5,3), (5,2), (5,1) and (5,0).

First, let us determine the explicit finite difference equation that governs the 16 red-colored interior nodes, namely (1,1), (1,2), (1,3), (1,4), (2,1), (2,2), (2,3), (2,4). (3,1), (3,2), (3,3), (3,4), (4,1), (4,2), (4,3), and (4,4). Only conduction heat transfer affects these interior nodes from adjoining nodes as shown in Figure 9-3 for node (1,1). Also, we will use the convention that conduction heat is always flowing into the center node from adjoining nodes.

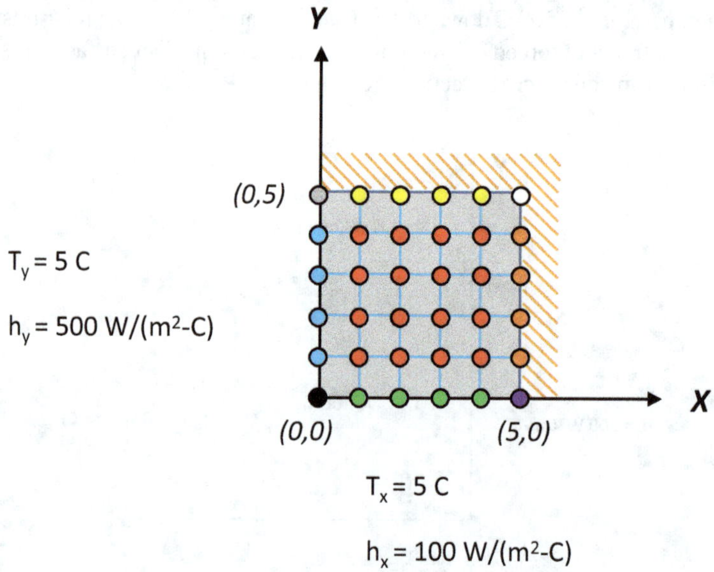

Figure 9-2 Explicit finite difference method is applied to one quarter of the square surface or to 36 node points where $\Delta X = \Delta Y = 0.1\ m$

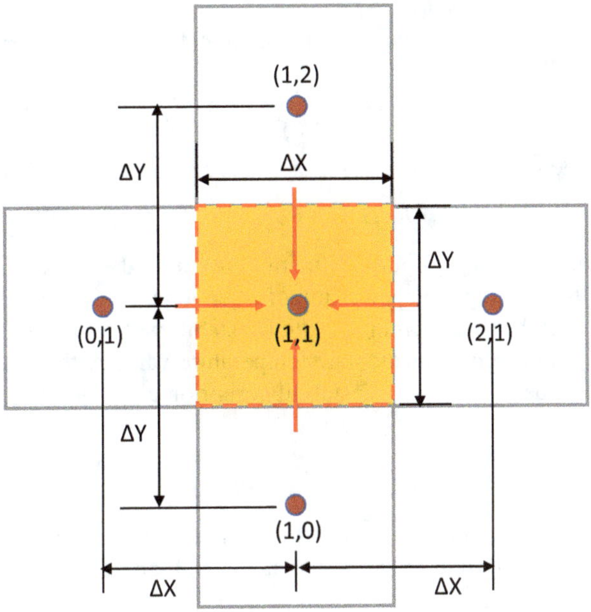

Figure 9-3 Conduction heat transfer to an interior node (1,1) from its adjoining nodes (1,0), (1,2), (0,1) and (2,1)

Conduction heat transfer equations to an interior node (1,1) from its adjoining nodes (1,0), (1,2), (0,1) and (2,1) at a previous time *j* are shown in Equations (9-2).

$$q_{(0,1)\rightarrow(1,1)} = k\left(\Delta Y \times 1\right)\frac{T^{j}_{(0,1)} - T^{j}_{(1,1)}}{\Delta X} \qquad q_{(2,1)\rightarrow(1,1)} = k\left(\Delta Y \times 1\right)\frac{T^{j}_{(2,1)} - T^{j}_{(1,1)}}{\Delta X} \tag{9-2}$$

$$q_{(1,0)\rightarrow(1,1)} = k\left(\Delta X \times 1\right)\frac{T^{j}_{(1,0)} - T^{j}_{(1,1)}}{\Delta Y} \qquad q_{(1,2)\rightarrow(1,1)} = k\left(\Delta X \times 1\right)\frac{T^{j}_{(1,2)} - T^{j}_{(1,1)}}{\Delta Y}$$

Letting $\Delta X = \Delta Y$, transient energy balance for node (1,1) is shown in Equation 9-3 which shows the internal thermal energy change of node (1,1) where superscript j denotes the previous time and $j + 1$ next time after a time increment of $\Delta\theta$ has elapsed.

$$T^{j}_{(0,1)} + T^{j}_{(2,1)} + T^{j}_{(1,0)} + T^{j}_{(1,2)} - 4 \times T^{j}_{(1,1)} = \frac{\rho C_p \left(\Delta X \times \Delta Y \times 1\right)}{k}\frac{\left(T^{j+1}_{(1,1)} - T^{j}_{(1,1)}\right)}{\Delta\theta} \tag{9-3}$$

Let $a = \frac{k}{\rho C_p}$ which is the constant thermal diffusivity of the material in m²/s and let the Fourier number $Fo = \frac{a\,\Delta\theta}{(\Delta X)^2}$, then $T^{j+1}_{(1,1)}$ can be determined from previous time's temperature values of $T^{j}_{(1,1)}$ and adjoining nodes as shown in Equation 9-4.

$$T^{j+1}_{(1,1)} = (1-4Fo)T^{j}_{(1,1)} + Fo\left(T^{j}_{(0,1)} + T^{j}_{(2,1)} + T^{j}_{(1,0)} + T^{j}_{(1,2)}\right) \qquad 9\text{-}4$$

The explicit finite difference Equation 9-4 is used 16 interior nodes shown in Figure 9-2. During the application of Equation 9-4, we should pay attention to the value of the time increment, $\Delta\theta$, for stability of results. For stable results which will approach a steady-state temperature value without oscillations for 16 interior nodes, the following stability criterion given in Equation 9-5 should be followed.

$$(1-4Fo) \geq 0 \quad \text{or} \quad \Delta\theta \leq 0.25\frac{(\Delta X)^2}{a} \qquad 9\text{-}5$$

Now, let us determine the explicit finite difference equation for the black-colored corner node (0,0) in Figure 9-2 as shown in Figure 9-4.

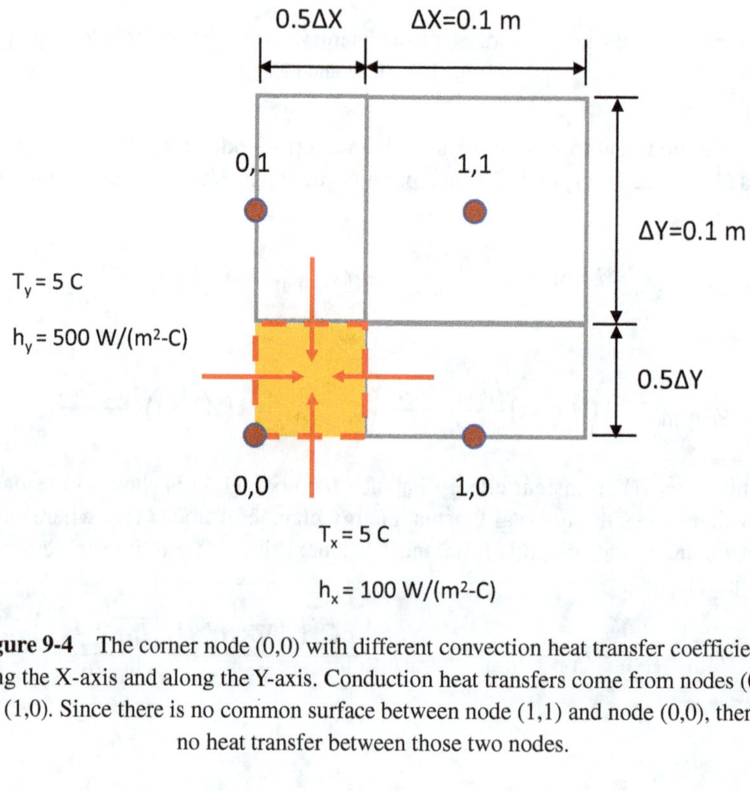

Figure 9-4 The corner node (0,0) with different convection heat transfer coefficients along the X-axis and along the Y-axis. Conduction heat transfers come from nodes (0,1) and (1,0). Since there is no common surface between node (1,1) and node (0,0), there is no heat transfer between those two nodes.

Node (0,0) is affected by different convection heat transfers along the X-axis and along the Y-axis and they are shown in Equations 9-6a and 9-6b.

$$q_{conv-y} = h_y\left(\frac{\Delta Y}{2}\times 1\right)\left(T_{\infty,y} - T^j_{(0,0)}\right) \quad \text{9-6a}$$

$$q_{conv-x} = h_x\left(\frac{\Delta X}{2}\times 1\right)\left(T_{\infty,x} - T^j_{(0,0)}\right) \quad \text{9-6b}$$

The transient energy balance for node (0,0) is shown in Equation 9-7. Node (0.0) is also affected by conduction heat transfer from nodes (0,1) and (1,0).

$$T^{j+1}_{(0,0)} = \left(1 - 2Bi_x Fo - 2Bi_y Fo - 4Fo\right)T^j_{(0,0)} +$$

$$2Bi_x FoT_{\infty,x} + 2Bi_y FoT_{\infty,y} + 2Fo\left(T^j_{(0,1)} + T^j_{(1,0)}\right) \quad \text{9-7}$$

The $\Delta\theta$ stability criterion for Equation 9-7 is provided in Equation 9-8.

$$1 - 2Bi_x Fo - 2Bi_y Fo - 4Fo \le 0 \text{ or } \Delta\theta \le \frac{0.5(\Delta X)^2}{\alpha\left(Bi_x + Bi_y + 2\right)} \quad \text{9-8}$$

where $Bi_x = \frac{h_x \Delta X}{k}$ and $Bi_y = \frac{h_y \Delta Y}{k}$ are respective finite difference form of Biot numbers along the X-axis and the Y-axis.

Next, let us determine the explicit finite difference equation for green-colored X-axis nodes (1,0), (2,0), (3,0) and (4,0) with convection heat transfer. Effect of convection heat transfer to node (1,0) is shown in Equation 9-9.

$$q_{conv-x} = h_x\left(\Delta X \times 1\right)\left(T_{\infty,x} - T^j_{(1,0)}\right) \quad \text{9-9}$$

Node (1,0) is also affected by conduction heat transfer from adjoining nodes (0,0), (2,0) and (1,1). The transient energy balance for node (1,0) is shown in Equation 9-10.

$$T^{j+1}_{(1,0)} = \left(1 - 2Bi_x Fo - 4Fo\right)T^j_{(1,0)} + 2Bi_x FoT_{\infty,x} + Fo\left(2T^j_{(1,1)} + T^j_{(0,0)} + T^j_{(2,0)}\right) \quad \text{9-10}$$

The $\Delta\theta$ stability criterion for Equation 9-10 is provided in Equation 9-11.

$$1 - 2Bi_x Fo - 4Fo \le 0 \text{ or } \Delta\theta \le \frac{0.5(\Delta X)^2}{\alpha\left(Bi_x + 2\right)} \quad \text{9-11}$$

Now, we will determine the explicit finite difference equation for blue-colored Y-axis nodes (0,1), (0,2), (0,3) and (0,4) with convection heat transfer. Effect of convection heat transfer to node (0,1) is shown in Equation 9-12.

$$q_{conv-y} = h_y\left(\Delta Y \times 1\right)\left(T_{\infty,y} - T^j_{(0,1)}\right) \quad \text{9-12}$$

Node (0,1) is also affected by conduction heat transfer from adjoining nodes (0,0), (0,2) and (1,1). The transient energy balance for node (0,1) is shown in Equation 9-13.

$$T_{(0,1)}^{j+1} = \left(1 - 2Bi_y Fo - 4Fo\right) T_{(0,1)}^{j} + 2Bi_y Fo T_{\infty,y} + Fo\left(2T_{(1,1)}^{j} + T_{(0,0)}^{j} + T_{(0,2)}^{j}\right) \quad 9\text{-}13$$

The $\Delta\theta$ stability criterion for Equation 9-13 is same as Equation 9-11's stability criterion and it is provided in Equation 9-14.

$$1 - 2Bi_y Fo - 4Fo \leq 0 \text{ or } \Delta\theta \leq \frac{0.5(\Delta Y)^2}{\alpha\left(Bi_y + 2\right)} \quad 9\text{-}14$$

Next, we will determine the explicit finite difference equation for gray-colored Y-axis node (0,5) with convection heat transfer from left side, insulated on top side and conduction heat transfer from right and bottom sides as shown in Figure 9-2. The transient energy balance for node (0,5) is shown in Equation 9-15.

$$T_{(0,5)}^{j+1} = \left(1 - 2Bi_y Fo - 4Fo\right) T_{(0,5)}^{j} + 2Bi_y Fo T_{\infty,y} + 2Fo\left(T_{(0,4)}^{j} + T_{(1,5)}^{j}\right) \quad 9\text{-}15$$

The $\Delta\theta$ stability criterion for Equation 9-15 is same as in Equation 9-14.

Next, we will determine the explicit finite difference equation for yellow-colored insulated nodes (1,5), (2,5), (3,5) and (4,5) which are insulated at their top side and experience conduction heat transfer from left side, right side and bottom side as shown in Figure 9-2. The transient energy balance for node (1,5) is shown in Equation 9-16.

$$T_{(1,5)}^{j+1} = \left(1 - 4Fo\right) T_{(1,5)}^{j} + Fo\left(2T_{(1,4)}^{j} + T_{(0,5)}^{j} + T_{(2,5)}^{j}\right) \quad 9\text{-}16$$

The $\Delta\theta$ stability criterion for Equation 9-16 is same as in Equation 9-5.

Now, let us determine the explicit finite difference equation for white-colored corner node (5,5)-which is insulated at its top and at its right sides and it has conduction heat transfer from its left and bottom sides as shown in Figure 9-2. The transient energy balance for node (5,5) is shown in Equation 9-17.

$$T_{(5,5)}^{j+1} = \left(1 - 4Fo\right) T_{(5,5)}^{j} + 2Fo\left(T_{(4,5)}^{j} + T_{(5,4)}^{j}\right) \quad 9\text{-}17$$

The $\Delta\theta$ stability criterion for Equation 9-17 is same as in Equation 9-5.

Next, we will determine the explicit finite difference equation for orange-colored insulated nodes (5,1), (5,2), (5,3) and (5,4) which are insulated at their right side and experience conduction heat transfer from top, from bottom and left sides. The transient energy balance for node (5,1) is similar to node (1,5)'s transient energy balance which was presented in Equation 9-16. The transient energy balance for node (5,1) is shown in Equation 9-18.

$$T_{(5,1)}^{j+1} = \left(1 - 4Fo\right) T_{(5,1)}^{j} + Fo\left(2T_{(4,1)}^{j} + T_{(5,0)}^{j} + T_{(5,2)}^{j}\right) \quad 9\text{-}18$$

The $\Delta\theta$ stability criterion for Equation 9-18 is same as in Equation 9-5.

Lastly, we will determine the explicit finite difference equation for X-axis node (5,0) with convection heat transfer from bottom side, insulated on right side and conduction heat transfer from left and top sides as shown in Figure 9-2. The transient energy balance for node (5,0) is like the transient energy balance for node (0,5) shown in Equation 9-15. The transient energy balance for node (5,0) is shown in Equation 9-19.

$$T_{(5,0)}^{j+1} = \left(1 - 2Bi_x Fo - 4Fo\right) T_{(5,0)}^{j} + 2Bi_x Fo T_{\infty,x} + 2Fo\left(T_{(4,0)}^{j} + T_{(5,1)}^{j}\right) \quad 9\text{-}19$$

The $\Delta\theta$ stability criterion for Equation 9-19 is like the one provided in Equation 9-14 and it is given in Equation 9-20.

$$1 - 2Bi_x Fo - 4Fo \le 0 \text{ or } \Delta\theta \le \frac{0.5\left(\Delta X\right)^2}{\alpha\left(Bi_x + 2\right)} \quad 9\text{-}20$$

Now, we have all explicit finite difference equations that cover the 36 nodes. However, we do not know how large of a $\Delta\theta$ we can take during our calculations so that our calculations are stable. Let us use constant physical properties for the stainless steel at the film temperature of $T_{ss-film} = 0.5\left(500 + 5\right) = 252.5\ C$.

Stainless steel thermal conductivity: $k = 17\dfrac{W}{m - C}$

Stainless steel thermal diffusivity: $\alpha = 4.17 \times 10^{-6}\dfrac{m^2}{s}$

Then, Biot numbers are: $Bi_x = \dfrac{h_x \Delta X}{k} = 0.58$ and $Bi_y = \dfrac{h_y \Delta Y}{k} = 2.94$

We also have $\Delta X = \Delta Y = 0.1\ m$.

For 16 interior nodes, and for 9 insulated nodes with no convection heat transfer, for stable explicit finite difference calculations, the Fourier number $FO \le 0.25$ as shown in Equation 9-21.

$$Fo = \frac{\alpha \times \Delta\theta}{\left(\Delta X\right)^2} \le 0.25 \text{ or } \Delta\theta \le \frac{0.25\left(\Delta X\right)^2}{\alpha} = 600\ s \quad 9\text{-}21$$

For node (0,0), $Fo \le \dfrac{1}{2\left(Bi_x + Bi_y + 2\right)}$ as shown in Equation 9-22.

$$Fo \le \frac{1}{2\left(Bi_x + Bi_y + 2\right)} \text{ or } \Delta\theta \le \frac{0.5\left(\Delta X\right)^2}{\alpha\left(Bi_x + Bi_y + 2\right)} = 217\ \mathbf{s} \quad 9\text{-}22$$

For nodes (1,0), (2,0), (3,0), (4,0) and (5,0), $Fo \le \dfrac{1}{2\left(Bi_x + 2\right)}$ as shown in Equation 9-23.

$$Fo \le \frac{1}{2\left(Bi_x + 2\right)} \text{ or } \Delta\theta \le \frac{0.5\left(\Delta X\right)^2}{\alpha\left(Bi_x + 2\right)} = 464\ s \quad 9\text{-}23$$

For nodes (0,1), (0,2), (0,3), (0,4) and (0,5), $Fo \le \dfrac{1}{2(Bi_y + 2)}$ as shown in Equation 9-24.

$$Fo \le \frac{1}{2(Bi_y + 2)} \quad \text{or} \quad \Delta\theta \le \frac{0.5(\Delta Y)^2}{\alpha\,(Bi_y + 2)} = 243\,s \qquad \text{9-24}$$

For stability of our calculations, the most stringent $\Delta\theta$ is at the corner node (0,0), namely $\Delta\theta \le 217\ s$. For the present calculations a $\Delta\theta$ of 200 seconds is used.

36 explicit finite difference equations presented above are solved using an MS Excel spreadsheet. Time is increased with 200-second intervals until the mid-point of the stainless-steel rectangular prism bar's square surface, node (5,5), cooled down from 500 C to 100 C. This cooling curve is shown in Figure 9-5. With the specified convection heat transfer coefficients in Figure 9-1, it takes 9.64 hours to cool the mid-point of the stainless-steel rectangular prism bar's square surface from 500 C to 100 C.

Temperature goes down faster at other nodes as expected. As expected, fastest cooling of the stainless-steel rectangular prism bar's square surface occurs at node (0,0) as shown in Figure 9-6. At location X=0.25 m and Y=0.25 m, the average temperature of four surrounding nodes (2,2), (2,3), (3,2) and (3,3) was used.

Constant temperature curves, isotherms, after 1 hour of cooling are shown in Figure 9-7 for 200 C, 300 C, 400 C and 450 C. These constant temperature

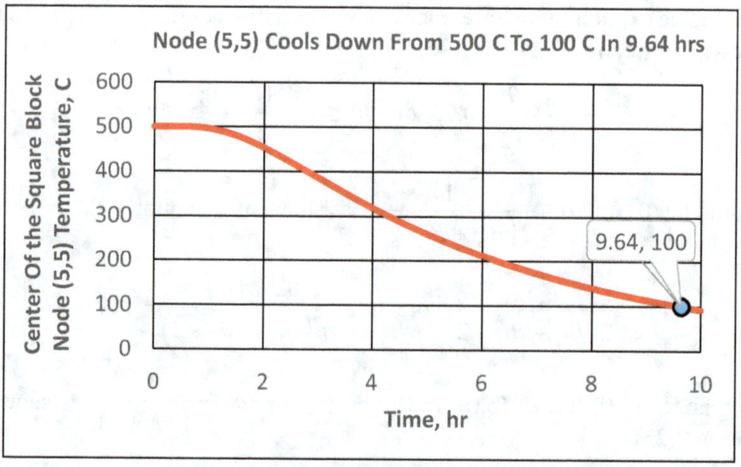

Figure 9-5 Temperature versus time curve for node (5,5), namely the mid-point of the stainless-steel rectangular prism bar's square surface

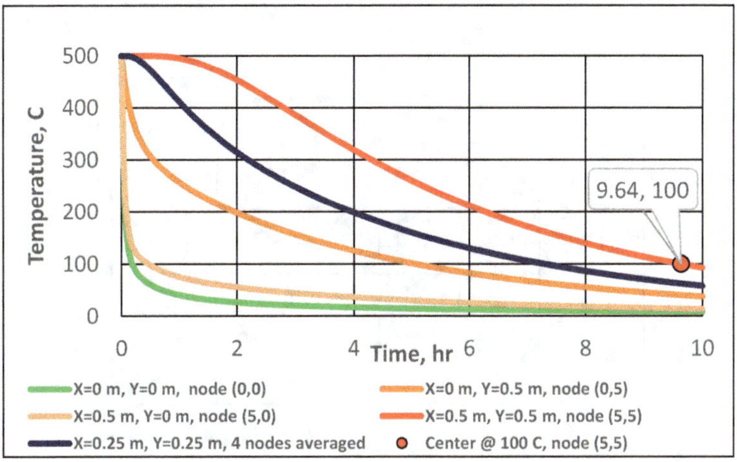

Figure 9-6 Temperature versus time curves for different nodes

curves are obtained by interpolation of temperature values at a node and its adjoining node. Temperature values that are shown in Figure 9-7 are carried to other three quarters of the square surface by symmetry. Figure 9-8 shows constant temperature curves for the whole square surface after 1 hour of cooling. Constant temperature curves spread out more in the X-axis direction than the spread in

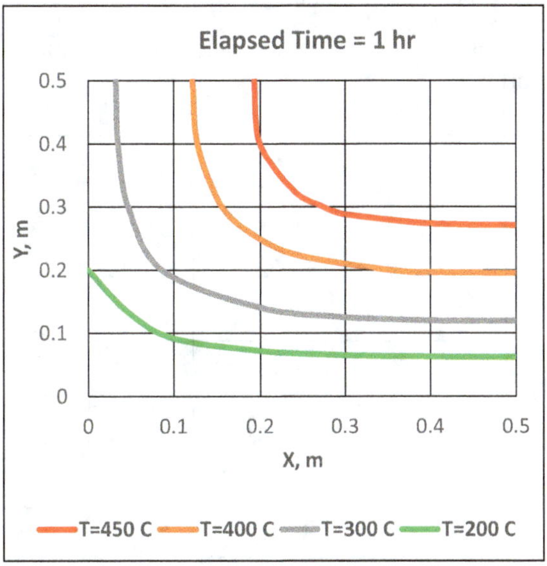

Figure 9-7 Isotherms at one quarter of the square surface obtained by interpolation between nodes after 1 hour of cooling

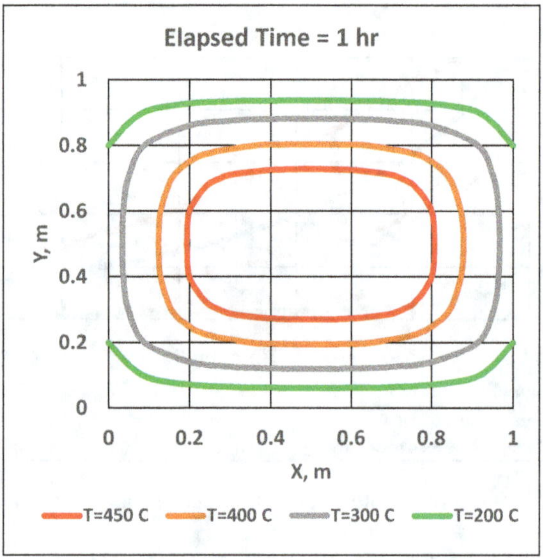

Figure 9-8 Isotherms for the whole square surface after 1 hour of cooling obtained from Figure 9-7 temperature values using symmetry

Y-axis direction. This is due to higher convection heat transfer coefficient on the Y-axis surface, $h_y = 500\frac{W}{m^2-C}$, than the X-axis surface, $h_x = 100\frac{W}{m^2-C}$, with the same T_∞ at both surfaces which causes higher cooling from Y-axis surfaces than the X-axis surfaces.

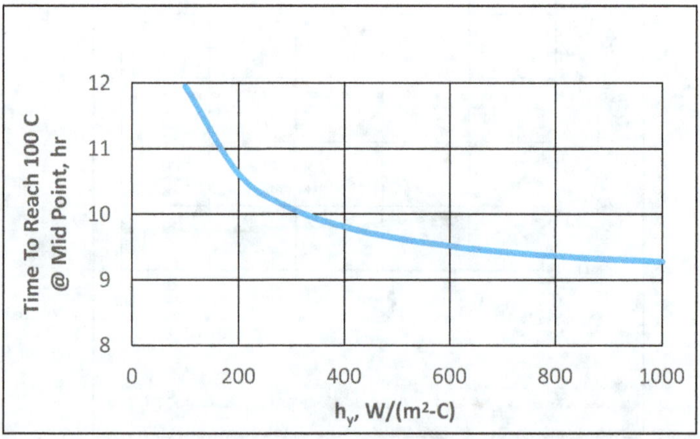

Figure 9-9 Cooling times from 500 C to 100 C at node (5,5) versus convection heat transfer coefficient at Y-axis surfaces

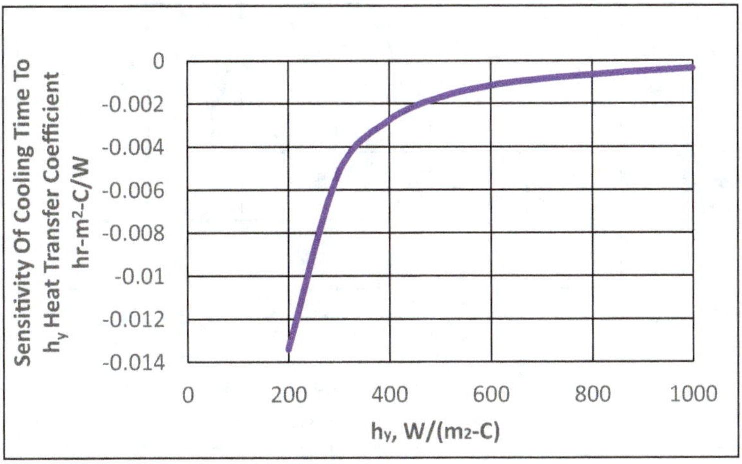

Figure 9-10 Sensitivity of cooling times from 500 C to 100 C at node (5,5) to changes in convection heat transfer coefficient at Y-axis surfaces

Figure 9-9 shows cooling times from 500 C to 100 C at node (5,5) versus changing convection heat transfer coefficient at Y-axis surfaces while every other property being the same. Cooling times decrease fast as the convection heat transfer coefficient at Y-axis surfaces increases from 100 W/(m^2-C) to 400 W/(m^2-C). Decrease in cooling times slows down as the convection heat transfer coefficient at Y-axis surfaces increases from 600 W/(m^2-C) to 1,000 W/(m^2-C). The sensitivity of cooling times from 500 C to 100 C at node (5,5) to h_y, namely

$$d\left(Cooling\,Time\,from\,500C\,to\,100C\,at\,node(5,5)\right)/\,dh_y,$$

is shown in Figure 9-10. At low values of h_y, around 200 W/(m^2-C), this sensitivity is -48.3 (s-m^2-C)/W. At high values of h_y, around 1,000 W/(m^2-C), this sensitivity is only -1.3 (s-m^2-C)/W.

Figure 9-11 shows cooling times from 500 C to 100 C at node (5,5) versus changing convection heat transfer coefficient at X-axis surfaces while every other property being the same. Cooling times decrease fast as the convection heat transfer coefficient at X-axis surfaces increases from 25 W/(m^2-C) to 100 W/(m^2-C). Decrease in cooling times slows down as the convection heat transfer coefficient at X-axis surfaces increases from 200 W/(m^2-C) to 300 W/(m^2-C). The sensitivity of cooling times from 500 C to 100 C at node (5,5) to h_x, namely

$$d\left(Cooling\,Time\,from\,500C\,to\,100C\,at\,node(5,5)\right)/\,dh_x,$$

is shown in Figure 9-12. At low values of h_x, around 50 W/(m^2-C), the cooling time sensitivity is -172 (s-m^2-C)/W. At high values of h_x, around 300 W/(m^2-C), the cooling time sensitivity is reduces to -15 (s-m^2-C)/W.

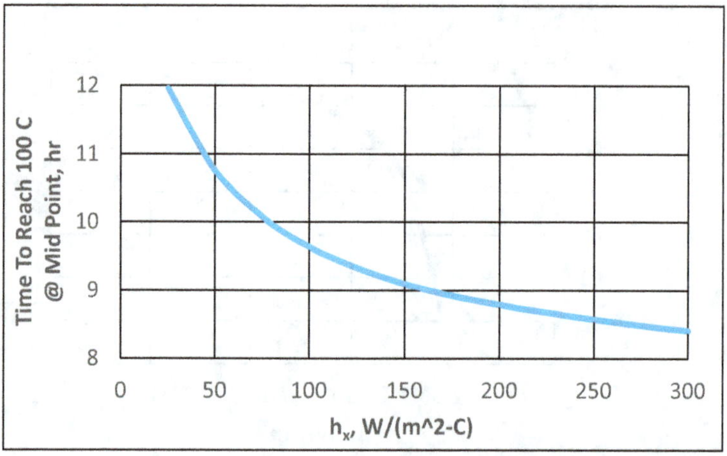

Figure 9-11 Cooling times from 500 C to 100 C at node (5,5) versus convection heat transfer coefficient at X-axis surfaces

Also, the sensitivity of cooling times from 500 C to 100 C at node (5,5) to changes in environment's temperature, T_∞, from 5 C to 20 was evaluated while every other property being the same, namely

$$\mathrm{d}\left(Cooling\,Time\,from\,500C\,to\,100C\,at\,node(5,5)\right)/\,dT_\infty\,.$$

This sensitivity of cooling times from 500 C to 100 C at node (5,5) to changes in environment's temperature, T_∞, from 5 C to 20 C was almost a constant at 158

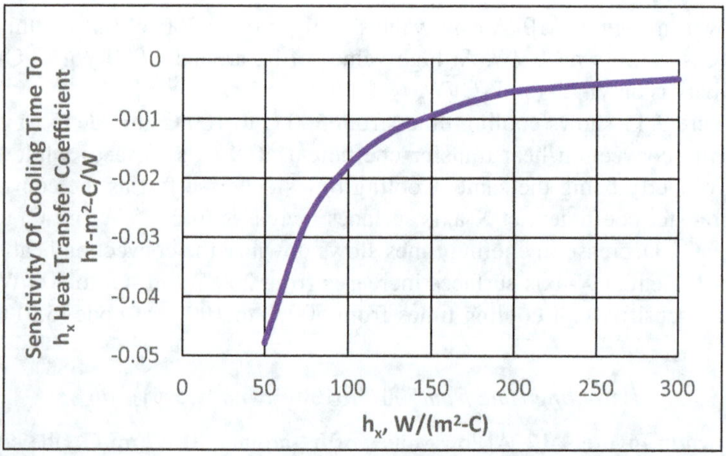

Figure 9-12 Sensitivity of cooling time s from 500 C to 100 C at node (5,5) to changes in convection heat transfer coefficient at X-axis surfaces

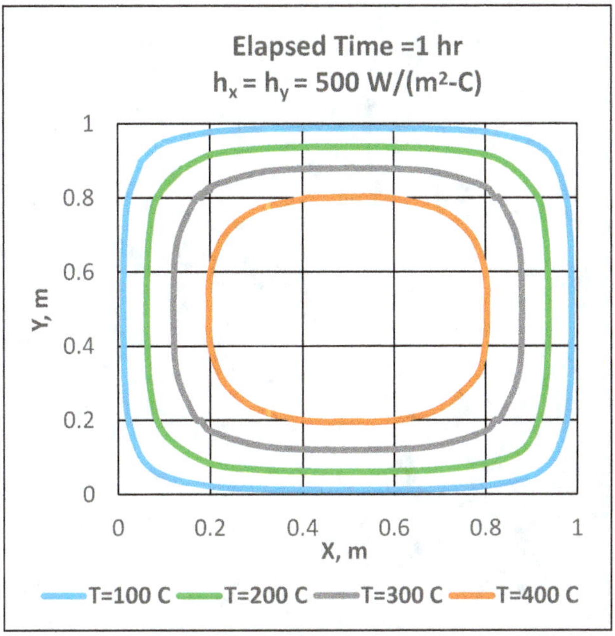

Figure 9-13 Isotherms for $h_x = h_y = 500\dfrac{W}{\left(m^2 - C\right)}$

seconds increase in cooling time per 1 C increase in environment's temperature. In the present analysis, cooling times from 500 C to 100 C at node (5,5) is more sensitive to changes in h_x than to changes in h_y. This is due to higher energy availability in the X-axis surfaces for heat transfer which can be seen from ellipse shaped constant temperature contours whose major axes are in the X-axis direction, see in Figure 9-8. As values of h_x get closer to values of h_y, these constant temperature contours will approach to shape of a square as shown in Figure 9-13 for $\theta = 1\ hr$. Then, the sensitivity to changes in h_x and the sensitivity to changes in h_y will equalize while every other property being the same.

10

A Triangular Prism whose Boundaries are Subject to Radiation Heat Transfer

In this chapter, we will cover the explicit finite difference method application to a long triangular prism with radiation heat transfer at its boundaries in a vacuum environment in transient and 2-dimensional mode. Let us investigate the heating of this titanium alloy triangular prism, with insulated ends from an initial temperature of 20 C (293 K) up to 600 C (873 K) at its center of gravity. The triangle is an equilateral one with 1 m sides. Three sides of the long triangular prism are subject to radiation heat transfer from radiation energy sources at 1,000 C (1,273 K) in a vacuum environment as shown in end view in Figure 10-1.

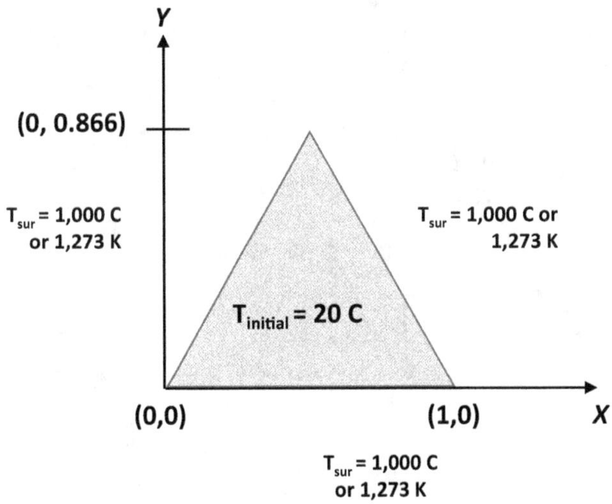

Figure 10-1 Cross-section of a titanium alloy long triangular prism subject to radiation heat transfer at its three boundaries

The 2-dimensional transient heat transfer equation with constant properties and with no internal energy generation is shown in Equation 10-1.

$$\frac{\partial^2 T}{\partial X^2}+\frac{\partial^2 T}{\partial Y^2}=\frac{1}{\alpha}\frac{\partial T}{\partial \theta} \quad \text{10-1}$$

where $T(X, Y)$ is the temperature on the cross-section of the titanium alloy long triangular prism at (X, Y) location in K. θ is time in s. α is titanium alloy's thermal diffusivity in m²/s. Explicit finite difference method is applied to Equation 10-1 by using conservation of energy at each (X, Y) node. Since there is symmetry about X = 0.5 m line in radiation heat transfer affecting three boundaries of the triangular prism, only half of the (X, Y) surface about the X = 0.5 m line will be analyzed as shown in Figure 10-2.

In Figure 10-2, orange dash lines represent the symmetry boundary at X = 0.5 m and nodes on these boundaries will be treated as insulated nodes on their right, namely nodes (5,0), (5,1), (5,2), (5,3), (5,4), and (5,5).

First, the explicit finite difference equation that governs 6 interior red-colored nodes, namely nodes (2,1), (3,1), (4,1), (3,2), (4,2), and (4,3), is the similar to the one in Chapter 9, Equation 9.4 except that $\Delta X \neq \Delta Y$, namely in the present case $\Delta X = 0.1m$ and $\Delta Y = 0.173m$, as shown in Equation 10-2 for node (3,1). Only conduction heat transfer affects these interior nodes from adjoining nodes. Also, we use the convention that conduction heat is always flowing into the center node from adjoining nodes.

$$T_{(3,1)}^{j+1}=T_{(3,1)}^{j}+Fo_{\Delta X}\left(T_{(2,1)}^{j}+T_{(4,1)}^{j}-2T_{(3,1)}^{j}\right)+Fo_{\Delta Y}\left(T_{(3,0)}^{j}+T_{(3,2)}^{j}-2T_{(3,1)}^{j}\right) \quad \text{10-2}$$

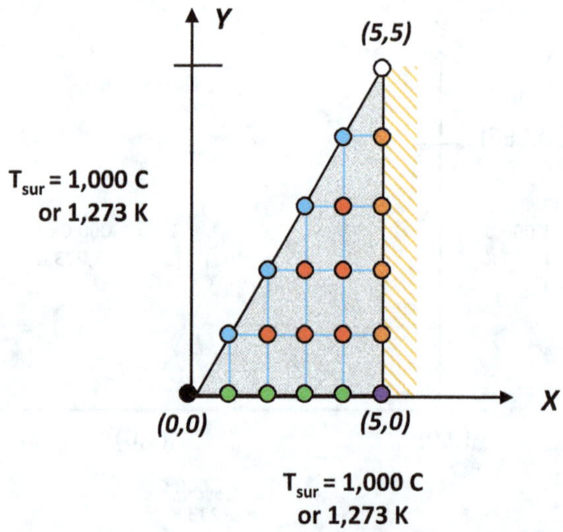

Figure 10-2 Explicit finite difference method is applied to half of the equilateral triangle's surface with 21 node points where $\Delta X = 0.1\ m$ and $\Delta Y = 0.173\ m$

where $Fo_{\Delta X} = \frac{a\,\Delta\theta}{(\Delta X)^2}$ and $Fo_{\Delta Y} = \frac{a\,\Delta\theta}{(\Delta Y)^2}$. The stability criterion for solutions of Equation 10-2 is given in Equation 10-3.

$$1 - 2Fo_{\Delta X} - 2Fo_{\Delta Y} \geq 0 \qquad 10\text{-}3$$

Next, we will use the explicit finite difference equation for 4 insulated orange-colored nodes, namely nodes (5,1), (5,2), (5,3) and (5,4), which are insulated at their right side and experience conduction heat transfer from top, bottom and left adjoining nodes. The transient energy balance for node (5,1) is similar to the one in Chapter 9, Equation 9.18 except that $\Delta X \neq \Delta Y$, namely in the present case $\Delta X = 0.1\,m$ and $\Delta Y = 0.173\,m$, as shown in Equation 10-4 for node (5,1).

$$T_{(5,1)}^{j+1} = T_{(5,1)}^{j} + 4Fo_{\Delta X}\left(T_{(4,1)}^{j} - T_{(5,1)}^{j}\right) + Fo_{\Delta Y}\left(T_{(5,0)}^{j} + T_{5,2}^{j} - 2T_{(5,1)}^{j}\right) \qquad 10\text{-}4$$

The $\Delta\theta$ stability criterion for Equation 10-4 is given in Equation 10-5.

$$1 - 4Fo_{\Delta X} - 2Fo_{\Delta Y} \geq 0 \qquad 10\text{-}5$$

Next, let us determine the explicit finite difference equation for green-colored nodes (1,0), (2,0), (3,0) and (4,0) on X-axis with radiation heat transfer at the $Y = 0$ line boundary. Effect of radiation heat transfer to node (1,0) is shown in Equation 10-6.

$$q_{rad} = h_{rad}\left(\Delta X \times 1\right)\left(T_{sur} - T_{(1,0)}^{j}\right) = \varepsilon\sigma\left(\Delta X \times 1\right)\left(T_{sur}^{4} - T_{(1,0)}^{4,j}\right) \qquad 10\text{-}6$$

The radiation heat transfer coefficient at the surface is given in Equation 10-7.

$$h_{rad} = \varepsilon\sigma\left(T_{sur} + T_{(1,0)}^{j}\right)\left(T_{sur}^{2} + T_{(1,0)}^{2,j}\right) \qquad 10\text{-}7$$

where ε is the emissivity of the titanium surface which is assumed to be diffuse and gray and σ is the Stefan-Boltzmann constant at 5.67×10^{-8} W/(m^2-K^4). Node (1,0) is also affected by conduction heat transfer from adjoining nodes (0,0), (2,0) and (1,1). The transient energy balance for node (1,0) is shown in Equation 10-8.

$$T_{(1,0)}^{j+1} = T_{(1,0)}^{j} + 2Bi_{\Delta X}Fo_{\Delta X\Delta Y}\left(T_{sur} - T_{(1,0)}^{j}\right) + Fo_{\Delta X}\left(T_{(0,0)}^{j} + T_{(2,0)}^{j} - 2T_{(1,0)}^{j}\right) \qquad 10\text{-}8$$
$$+ 2Fo_{\Delta Y}\left(T_{(1,1)}^{j} - T_{(1,0)}^{j}\right)$$

where $Bi_{\Delta X} = \frac{h_{rad}\Delta X}{k}$ and $Fo_{\Delta X\Delta Y} = \frac{a\,\Delta\theta}{\Delta X\,\Delta Y}$. The $\Delta\theta$ stability criterion for Equation 10-8 is provided in Equation 10-9.

$$1 - 2Bi_{\Delta x}Fo_{\Delta X\Delta Y} - 2Fo_{\Delta X} - 2Fo_{\Delta Y} \leq 0 \qquad 10\text{-}9$$

Now, let us determine the explicit finite difference equation for blue-colored nodes (1,1), (2,2), (3,3) and (4,4) with radiation heat transfer at the line boundary. For node (1,1), the explicit finite difference equation is shown in Equation 10-10.

$$T_{(1,1)}^{j+1} = T_{(1,1)}^{j} + 2Bi_{0.2}Fo_{\Delta X\Delta Y}\left(T_{sur} - T_{(1,1)}^{j}\right) + 2Fo_{\Delta X}\left(T_{(2,1)}^{j} - T_{(1,1)}^{j}\right) + 2Fo_{\Delta Y}\left(T_{(1,0)}^{j} - T_{(1,1)}^{j}\right) \qquad 10\text{-}10$$

The $\Delta\theta$ stability criterion for Equation 10-10 is provided in Equation 10-11.

$$1 - 2Bi_{0.2}Fo_{\Delta X\Delta Y} - 2Fo_{\Delta X} - 2Fo_{\Delta Y} \leq 0 \qquad 10\text{-}11$$

where $Bi_{0.2} = \frac{h_{rad}0.2}{k}$ and k is the thermal conductivity of titanium in W/(m-K).

The black-colored corner node, point (0,0), has the following explicit finite difference equation given in Equation 10-12.

$$T_{(0,0)}^{j+1} = T_{(0,0)}^{j} + 2Bi_{\Delta X}Fo_{\Delta X\Delta Y}\left(T_{sur} - T_{(0,0)}^{j}\right) + 4Bi_{0.1}Fo_{\Delta X\Delta Y}\left(T_{sur} - T_{(0,0)}^{j}\right) + 2Fo_{\Delta X}\left(T_{(1,0)}^{j} - T_{(0,0)}^{j}\right) \qquad 10\text{-}12$$

where $Bi_{0.1} = \frac{h_{rad}0.1}{k}$. The $\Delta\theta$ stability criterion for Equation 10-12 is provided in Equation 10-13.

$$1 - 2Bi_{\Delta X}Fo_{\Delta X\Delta Y} - 4Bi_{0.1}Fo_{\Delta X\Delta Y} - 2Fo_{\Delta X} \leq 0 \qquad 10\text{-}13$$

The violet-colored corner node, point (5,0), has the following explicit finite difference equation given in Equation 10-14.

$$T_{(5,0)}^{j+1} = T_{(5,0)}^{j} + 2Bi_{\Delta X}Fo_{\Delta X\Delta Y}\left(T_{sur} - T_{(0,0)}^{j}\right) + 2Fo_{\Delta X}\left(T_{(4,0)}^{j} - T_{(0,0)}^{j}\right) + 2Fo_{\Delta Y}\left(T_{(5,1)}^{j} - T_{(0,0)}^{j}\right) \qquad 10\text{-}14$$

The $\Delta\theta$ stability criterion for Equation 10-14 is provided in Equation 10-15.

$$1 - 2Bi_{\Delta X}Fo_{\Delta X\Delta Y} - 2Fo_{\Delta X} - 2Fo_{\Delta Y} \leq 0 \qquad 10\text{-}15$$

Now, let us determine the explicit finite difference equation for the white-colored corner node (5,5)-which is insulated at its right side and it has radiation heat transfer from its left side. The transient energy balance for node (5,5) is shown in Equation 10-16.

$$T_{(5,5)}^{j+1} = T_{(5,5)}^{j} + 4Bi_{0.1}Fo_{\Delta X\Delta Y}\left(T_{sur} - T_{(5,5)}^{j}\right) + 2Fo_{\Delta Y}\left(T_{(5,4)}^{j} - T_{(5,5)}^{j}\right) \qquad 10\text{-}16$$

The $\Delta\theta$ stability criterion for Equation 10-16 is shown in Equation 10-17.

$$1 - 4Bi_{0.1}Fo_{\Delta X\Delta Y} - 2Fo_{\Delta Y} \leq 0 \qquad 10\text{-}17$$

Now, we have all explicit finite difference equations that cover 21 nodes. However, we do not know how large of a $\Delta\theta$ we can take during our calculations so that our calculations are stable. Let us use constant physical properties for the titanium material at its film temperature of

$$T_{titanium-film} = 0.5(20+600)+273 = 583\,K.$$

Titanium thermal conductivity: $k = 20\dfrac{W}{m-K}$.

Titanium thermal diffusivity: $\alpha = 7.3\times10^{-6}\dfrac{m^2}{s}$.

We also have $\Delta X = 0.1m$ and $\Delta Y = 0.173m$

For 6 interior nodes with no convection heat transfer, for stable explicit finite difference calculations, Equation 10-3 has to be satisfied as shown in Equation 10-18.

$$\Delta\theta \le \frac{0.5}{\alpha} / \left[\frac{1}{(\Delta X)^2} + \frac{1}{(\Delta Y)^2}\right] \text{ or } \Delta\theta \le 513.4\,s \qquad 10\text{-}18$$

For 4 insulated nodes with no convection heat transfer, for stable explicit finite difference calculations, Equation 10-5 has to be satisfied as shown in Equation 10-19.

$$\Delta\theta \le \frac{0.5}{\alpha} / \left[\frac{2}{(\Delta X)^2} + \frac{1}{(\Delta Y)^2}\right] \text{ or } \Delta\theta \le 293.4\,s \qquad 10\text{-}19$$

Next, let us determine the stability criteria for green-colored nodes (1,0), (2,0), (3,0) and (4,0) on X-axis with radiation heat transfer at the $Y = 0$ line boundary, Equation 10-9. For this stability criteria, let us use the maximum radiation heat transfer coefficient which occurs when surface temperature of the triangular prism reaches 1,273 K, namely $h_{rad} = 1{,}871.5\,W/(m^2-K)$. This stability criteria is in Equation 10-20.

$$\Delta\theta \le \frac{0.5}{\alpha} / \left[\frac{Bi_{\Delta X}}{\Delta X \Delta Y} + \frac{1}{(\Delta X)^2} + \frac{1}{(\Delta Y)^2}\right] \text{ or } \Delta\theta \le 365.3\,s \qquad 10\text{-}20$$

Now, let us determine the stability criteria for blue-colored nodes (1,1), (2,2), (3,3) and (4,4) with radiation heat transfer, Equation 10-11, at the $Y = 1.73X$ line boundary. This stability criteria is in Equation 10-21.

$$\Delta\theta \le \frac{0.5}{\alpha} / \left[\frac{Bi_{0.2}}{\Delta X \Delta Y} + \frac{1}{(\Delta X)^2} + \frac{1}{(\Delta Y)^2}\right] \text{ or } \Delta\theta \le 283.5\,s \qquad 10\text{-}21$$

The stability criteria given in Equation 10-13 for the black-colored corner node, point (0,0), is shown in Equation 10-22.

$$\Delta\theta \le \frac{0.5}{\alpha} / \left[\frac{Bi_{\Delta X}}{\Delta X \Delta Y} + \frac{2Bi_{0.1}}{\Delta X \Delta Y} + \frac{1}{(\Delta X)^2}\right] \text{ or } \Delta\theta \le 261.2\,s \qquad 10\text{-}22$$

The stability criteria given in Equation 10-15 for the violet-colored corner node, point (5,0), is shown in Equation 10-23.

$$\Delta\theta \le \frac{0.5}{a} / \left[\frac{Bi_{\Delta X}}{\Delta X \Delta Y} + \frac{1}{(\Delta X)^2} + \frac{1}{(\Delta Y)^2} \right] \text{ or } \Delta\theta \le 365.3\,s \qquad 10\text{-}23$$

The stability criteria given in Equation 10-17 for the white-colored corner node, point (5,5), is shown in Equation 10-24.

$$\Delta\theta \le \frac{0.5}{a} / \left[\frac{2Bi_{0.1}}{\Delta X \Delta Y} + \frac{1}{(\Delta Y)^2} \right] \text{ or } \Delta\theta \le 483.7\,s \qquad 10\text{-}24$$

The minimum $\Delta\theta$ criterion is at node (0,0) in Equation 10-22. *For the present explicit finite difference calculations, a $\Delta\theta$ of 200 seconds is chosen.*

First, let us investigate the behavior of radiation heat transfer coefficient by time given in Equation 10-7 for node (1,0). Figure 10-3 shows the increase of radiation heat transfer coefficients by time for nodes (1,0) and (5,5). The radiation heat transfer coefficient at node (5,5) which is at X = 0.5 m, Y = 0.866 m, increases faster and asymptotes earlier than at node (1,0), namely X = 0.1 m, Y = 0.0 m. This is due to slower increase of surface temperature at node (1,0) than at node (5,5) because node (1,0) distributes its thermal energy to lower temperature interior nodes much more easily. Every surface node in this case has different radiation heat transfer coefficient versus elapsed time behavior.

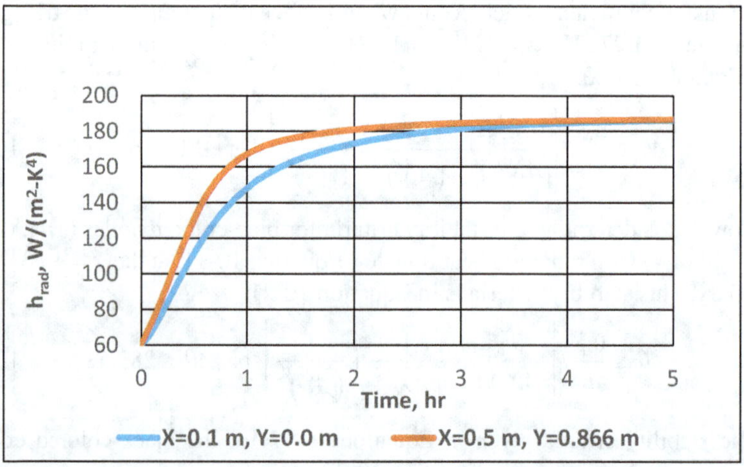

Figure 10-3 Radiation heat transfer coefficient versus time behavior for nodes (1,0) at X = 0.1m and Y=0.0 m and (5,5) at X = 0.5 m and Y = 0.866 m

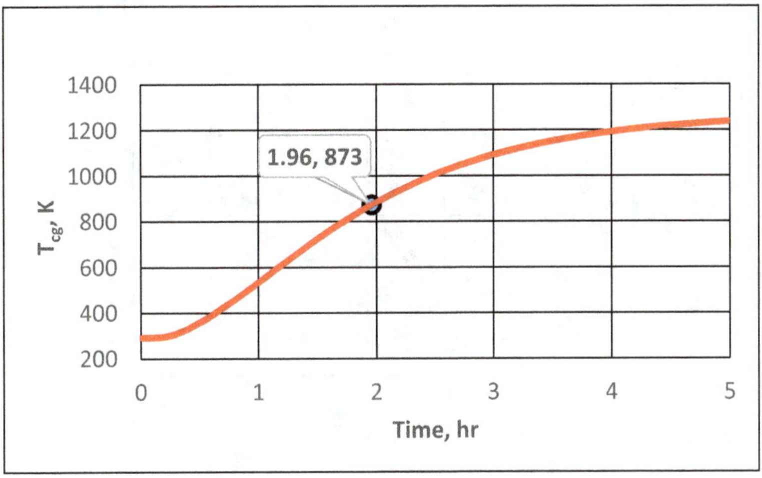

Figure 10-4 Temperature versus time curve for center-of-gravity of titanium triangular prism at X = 0.5 m and Y = 0.289 m

21 explicit finite difference equations presented above are solved using an MS Excel spreadsheet. Time is increased by 200-second intervals until the center-of-gravity point temperature of titanium triangular prism bar reaches 873 K. This increasing temperature versus time curve for X = 0.5 m and Y = 0.289 m is shown in Figure 10-4. The center-of-gravity temperature reaches to 873 K in 0.925 hour.

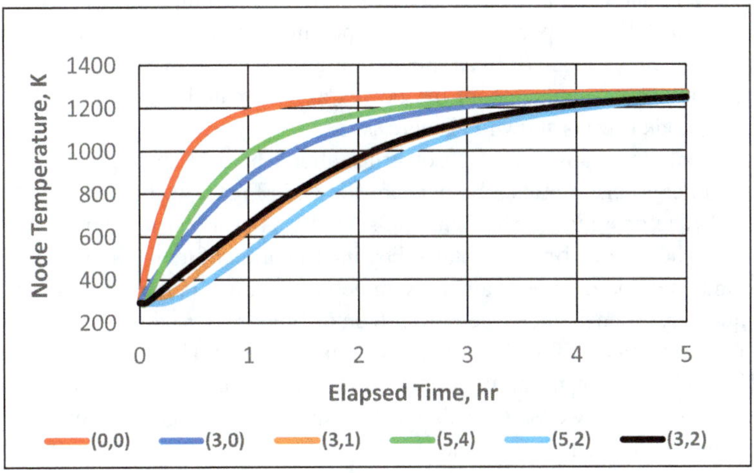

Figure 10-5 Different node temperatures versus elapsed time

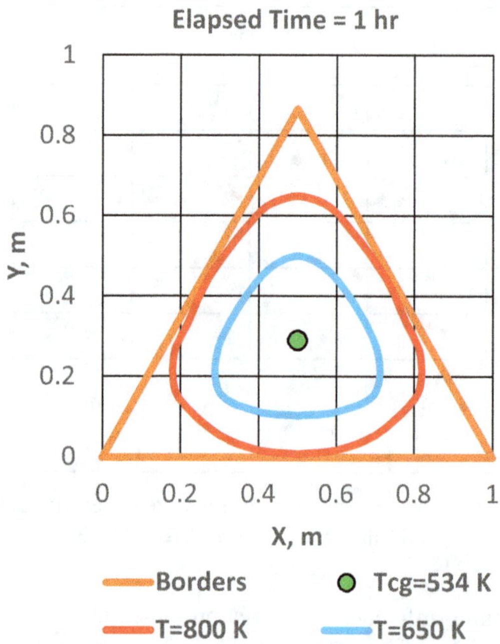

Figure 10-6 Isotherms at a cross-section of a long triangular prism after 1 hour of heating in vacuum by radiation heat transfer at its 3 borders

The center-of-gravity temperature at X = 0.5 m and Y = 0.289 m is obtained by linear interpolation between temperatures of nodes (5,1) at X = 0.5 m and Y = 0.173 m and (5,2) at X = 0.5 m and Y = 0.346 m. Temperatures go up faster at other locations as expected. Fastest temperature rise in the titanium triangular prism bar occurs at node (0,0) as shown in Figure 10-5. Slowest temperature rise in the titanium triangular prism bar occurs at center nodes as expected. As an example, node (5,2) is shown in Figure 10-5.

Constant temperature curves, isotherms, after 1 hour of heating in vacuum by radiation heat transfer at its 3 borders are shown in Figure 10-6 for 534 K, 650 K and 800 K temperatures. The radiation source was at a constant temperature of 1,273 K. These isotherms are obtained by interpolation of temperature values at a node and its adjoining nodes and by symmetry about $X = 0.5\ m$. Three corners of the equilateral triangle are heated faster than its other regions due to larger surface area for radiation heat transfer and less volume for heat transfer at these corners.

The emissivity of titanium surface in the present analysis is assumed to be diffuse and gray. However, the value of emissivity can change between 0.3 to 0.6 depending upon titanium's surface conditions. In all of our above calculations, $\varepsilon = 0.4$. Now, let us investigate the effect of changing emissivity on the elapsed time that it takes to heat the center-of-gravity of above triangular prism to 873 K.

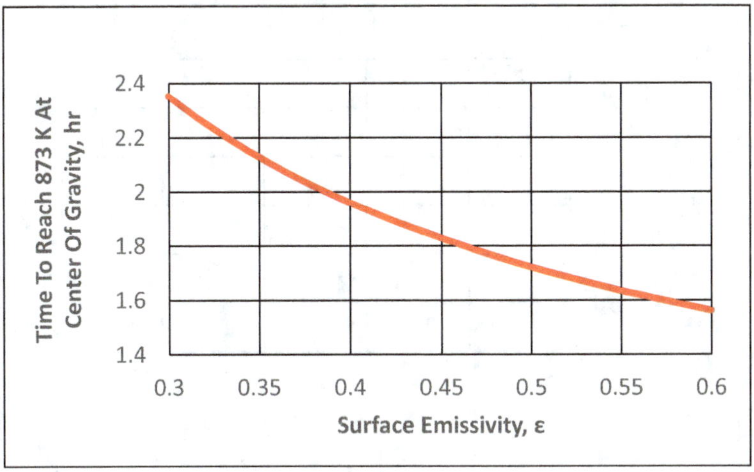

Figure 10-7 Time in hours to reach 873 K at center-of-gravity of above triangular prism versus its surface emissivity between 0.3 and 0.6

When $\varepsilon = 0.4$, it takes 1.96 hours to heat the center-of-gravity of above triangular prism from 293 K to 873 K while other thermophysical properties being constant. Time to reach 873 K at center-of-gravity of above triangular prism versus surface emissivity is shown in Figure 10-7. As the emissivity of the surface increases, time to reach 873 K at center of gravity decreases due to equal increase in surface absorptivity of radiation heat transfer.

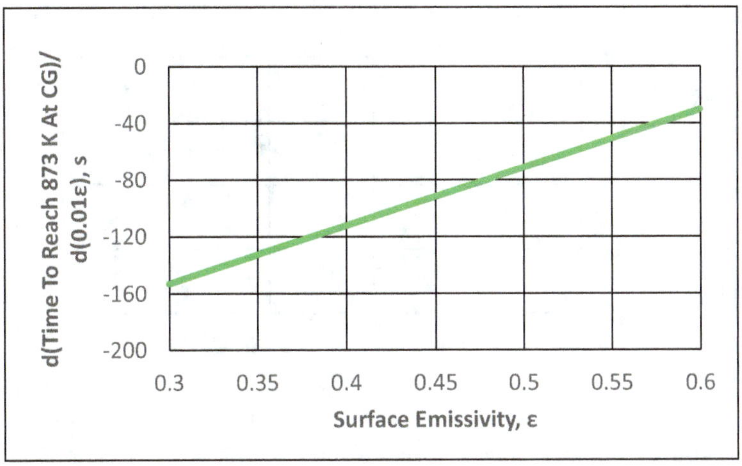

Figure 10-8 Sensitivity of time to reach 873 K at center-of-gravity of above triangular prism to 0.01 change in its surface emissivity

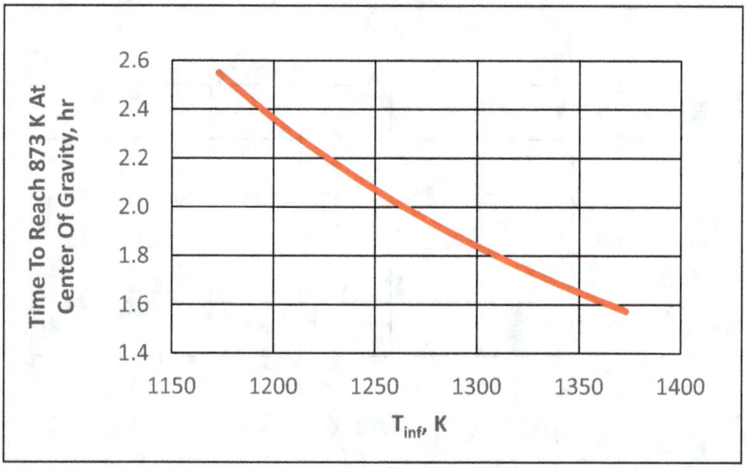

Figure 10-9 Time in hours to reach 873 K at center-of-gravity of above triangular prism versus its surrounding temperature between 1,173 K and 1,373 K

The sensitivity of time to reach 873 K at center-of-gravity of above triangular prism to its 0.01 change in its surface emissivity, namely $d(Time to reach\ 873\,K\ at\ center\ of\ gravity)/dT_{sur}$, is shown in Figure 10-8.

The absolute value of this sensitivity is high at low emissivity value of $\varepsilon = 0.3$. At $\varepsilon = 0.3$, a 153 s decrease in elapsed time is experienced to reach 873 K at the center of gravity when emissivity increases by 0.01. The absolute value of this

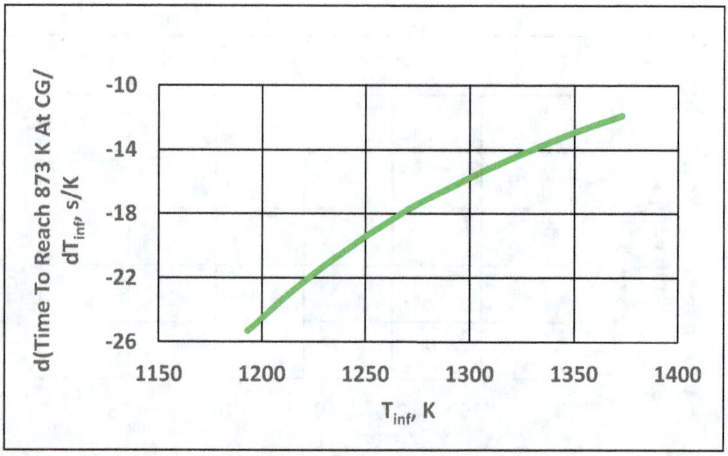

Figure 10-10 Sensitivity of time in seconds per K to reach 873 K at center-of-gravity of above triangular prism to changing surrounding temperature, T_{sur}.

sensitivity decreases at high emissivity values. When $\varepsilon = 0.6$, a 34 s decrease in elapsed time is experienced to reach 873 K at the center of gravity when emissivity increases by 0.01.

Another important variable in this analysis is the temperature of the radiation source, T_{sur}, which is at 1,273 K. As the surrounding temperature increases, elapsed time to reach 873K at center-of-gravity decreases as shown in Figure 10-9 for $T_{sur} = 1,273 \pm 100\,K$.

The sensitivity of time to reach 873 K at center-of-gravity of above triangular prism to changing surrounding temperature, namely $d\left(Time\,to\,reach\;873\,K\;at\;center\;of\;gravity\right)/\,dT_{sur}$, is shown in Figure 10-10.

Figure 10-10 Sensitivity of time in seconds per K to reach 873 K at center-of-gravity of above triangular prism to changing surrounding temperature,

This absolute value of this sensitivity is high at low surrounding temperature, i.e. 1,173 K, at 25.3 *s/K*. The absolute value of this sensitivity goes down at high surrounding temperature, i.e. 1,373 K, to 11.9 *s/K*.

11

Heat Treating a Long Cylindrical Pine Log in Vacuum

In this chapter, we will investigate heat treatment of a long cylindrical pine log, namely its length is at least ten times its radius, in vacuum. This kind of heat treatment of pine logs reduces its water content and therefore reduces its shrinkage, increases its dimensional stability and improves its decay while reducing its tensile and bending strengths. The pine log has a uniform initial temperature of 20 C. This pine log is treated like an infinite cylinder which suddenly experiences radiation heat transfer in a vacuum oven with a radiation source temperature of 200 C (473 K) and it has a radius, R_o, a length, L and $\frac{L}{R_o} > 10$. Transient temperature of this infinitely long pine log is given as a partial differential equation in cylindrical coordinates in Equation 11-1. See. Ref. [15].

$$\frac{\partial^2 T}{\partial R^2} + \frac{1}{R}\frac{\partial T}{\partial R} = \frac{1}{\alpha}\frac{\partial T}{\partial \theta} \qquad 11\text{-}1$$

The initial and two boundary conditions for Equation 11-1 are provided in Equations 11-2.

$$T(R,0) = T_{ini} \qquad 11\text{-}2a$$

$$\left(\frac{\partial T}{\partial R}\right)_{R=0} = 0 \qquad 11\text{-}2b$$

$$-k\left(\frac{\partial T}{\partial R}\right)_{R=R_o} = h_{rad}\left(T_\infty - T\right)_{R=R_o} \qquad 11\text{-}2c$$

where $T(R, \theta)$ is the temperature of the pine log at a radius R in meters and a time θ in seconds. α is the thermal diffusivity of pine log in m^2/s. k is the thermal conductivity of pine log in radial direction, perpendicular to grain direction, in W/(m-K). h_{rad} is the radiation heat transfer coefficient in W/(m^2-K) at pine log's surface given in Equation 11-3.

$$h_{rad} = \varepsilon\sigma\left(T_{film} + T_\infty\right)\left(T_{film}^2 + T_\infty^2\right) \qquad 11\text{-}3$$

where ε is the emissivity of pine log's surface which is assumed to be diffuse and gray and σ is the Stefan-Boltzmann constant at 5.67 × 10^{-8} W/(m^2-K^4).

T_{film} is the average temperature between pine log's initial uniform temperature and the surrounding temperature. The complete solution to Equations 1 and 2 is shown in Equation 11-4. See Ref. [15] for derivation of Equation 11-4.

$$\frac{T(R,\theta)-T_\infty}{T_{ini}-T_\infty}=\sum_{i=1}^{\infty}C_i e^{-\lambda_i^2 FO} J_0\left(\lambda_i \frac{R}{R_o}\right) \qquad 11\text{-}4$$

where FO is the Fourier number, the non-dimensional time, as shown in Equation 11-5.

$$FO=\frac{a\theta}{R_o^2} \qquad 11\text{-}5$$

Constants C_i is defined in Equation 11-6.

$$C_i=\left(\frac{2}{\lambda_i}\right)\frac{J_1(\lambda_i)}{J_o^2(\lambda_i)+J_1^2(\lambda_i)} \qquad 11\text{-}6$$

J_0 and J_1 are Bessel functions of the first kind. λ_i are positive roots of the transcendental Equation 11-7.

$$Bi=\lambda_i\frac{J_1(\lambda_i)}{J_0(\lambda_i)} \qquad 11\text{-}7$$

where Bi is the Biot number which is the ratio of the conduction heat transfer resistance in the pine log to surface radiation heat transfer resistance as shown in Equation 11-8.

$$Bi=\frac{R_o/k}{1/h_{rad}}=\frac{h_{rad}R_o}{k} \qquad 11\text{-}8$$

Let us apply a five-term expansion to Equation 11-4 and calculate the time it takes for the centerline of the pine log to reach 150 C (423 K) as shown in Equation 11-9 where $J_0(0) = 1$.

$$\frac{T(0,\theta)-T_\infty}{T_{ini}-T_\infty}\approx C_1e^{-\lambda_1^2 FO}+C_2e^{-\lambda_2^2 FO}+C_3e^{-\lambda_3^2 FO}+C_4e^{-\lambda_4^2 FO}+C_5e^{-\lambda_5^2 FO} \qquad 11\text{-}9$$

We will use the following constant properties for the pine log and its vacuum radiation environment to determine the time to time it takes for the centerline of the pine log to reach 150 C (423 K).

$T_{ini}=20\ C=293K;\ T_\infty=200\ C=473K$

$R_o=0.25m\ ;\ L=5m\ ;\ \frac{L}{R_o}=20>10$ Therefore, it is valid to use the transient and one-dimensional form of the heat transfer Equation 11-1 and its initial and boundary conditions, Equations 11-2, for an infinite cylinder to determine the time it takes for the center of the pine log to reach 150 C.

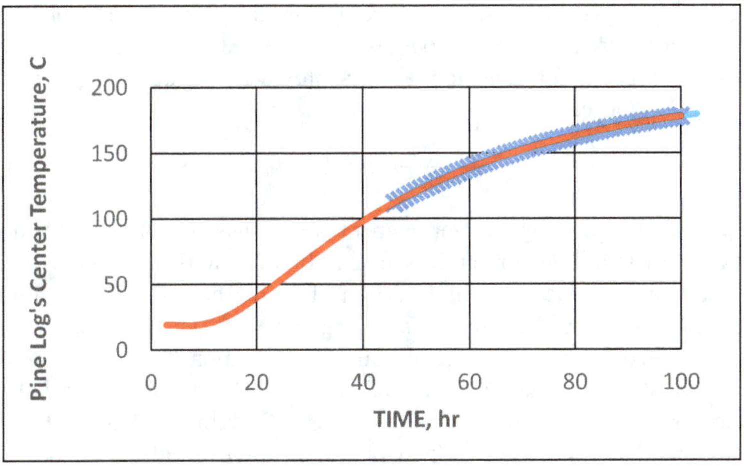

Figure 11-1 Pine log's center temperature versus time

$$k = 0.15\ W/(m-K) \quad ; \quad a = 8.36\times10^{-8}\ m^2/s$$

$$\varepsilon = 0.85 \quad ; \quad T_{film} = 0.5(20+200) = 110C = 383\,K$$

$$h_{rad} = 15.28\ W/(m^2-K)$$

All parameters that are required to solve Equation 11-9 are provided in the following Table 11-1.

Using all parameters in Table 11-1, we obtain the pine log's center temperature versus time shown as a red line in Figure 11-1. The center temperature of this pine log reaches from 20 C to 150 C in 68.074 hr or 245,068 s. The non-dimensional form of time to reach 150 C is $FO = 0.328$.

$Bi = 25.47$				
λ_1	λ_2	λ_3	λ_4	λ_5
2.3089	5.4029	8.3210	11.3522	14.3948
$J_0(\lambda_1)$	$J_0(\lambda_2)$	$J_0(\lambda_3)$	$J_0(\lambda_4)$	$J_0(\lambda_5)$
0.0508	-0.0403	0.0904	-0.1016	0.1074
$J_1(\lambda_1)$	$J_1(\lambda_2)$	$J_1(\lambda_3)$	$J_1(\lambda_4)$	$J_1(\lambda_4)$
0.5382	-0.3452	0.2668	-0.2185	0.1845
C_1	C_2	C_3	C_4	C_5
1.5954	-1.0580	0.8082	-0.6629	0.5624

Table 11-1 All required parameters are shown to 4 decimal places to solve Equation 11-9 for $Bi = 25.47$

Ref. [10] suggests that we can use only the first term in Equation 11-9, if $FO > 0.2$ to approximate Equation 11-4. For the centerline temperature, one term approximation for Equation 11-4 is provided in Equation 11-10, since $FO = 0.328$ in this case.

$$\frac{T(0,\theta)-T_\infty}{T_{ini}-T_\infty} \approx C_1 e^{-\lambda_1^2 FO} \qquad 11\text{-}10$$

We obtain the pine log's center temperature versus time for ε = 0.4 using the approximate 1-term Equation 11-10 shown as blue X's in Figure 11-1. The center temperature of this pine log using Equation 11-10 reaches from 20 C to 150 C in 68.084 hr or 245,102 s. Percentage error using only the 1st term in Equation 11-4 is only 0.014 as compared to the 5-term solution of Equation 11-9. It is important to remember that 1-term approximation to Equation 11-4 is appropriate, if $FO > 0.2$.

Another way to estimate the dimensionless temperature for transient conduction in an infinite cylinder, in an infinite plate or in a sphere with surface heat transfer is to use Heisler charts, see Ref. [15]. These charts provide the dimensionless temperature, $\frac{T(R,\theta)-T_\infty}{T_{ini}-T_\infty}$, as a function of Fourier number, $FO = \frac{a\theta}{R_o^2} > 0.2$, for various reciprocals of Biot number, $\frac{1}{Bi} = \frac{k}{h_{rad} R_o}$ or $\frac{k}{h_{rad} L}$.

For the present case, $\frac{T(0,\theta)-T_\infty}{T_{ini}-T_\infty} = \frac{(150-200)}{(20-200)} = 0.278$, $\frac{1}{Bi} = 0.0393$, and $\frac{R}{R_o} = 0$, and from Heisler charts in Ref. [15], FO is estimated to be 0.35. The present pine log using Heisler charts reaches from 20 C to 150 C in a 200 C radiation environment in 72.684 hr or 261,663 s. Percentage error using Heisler charts is 6.772 as compared to the 5-term solution of Equation 11-9.

For $FO = \frac{a\theta}{R_o^2} > 0.2$, the easiest way to determine the centerline temperature in an infinite cylinder with surface heat transfer is to use the one term approximation given in Equation 11-10 and using the graphs for λ_1 and C_1 given respectively, in Figures 11-2 and 11-3.

Now let us investigate the variation of pine log radius on time to reach 150 C at its center. In order to be able to use the infinite cylinder solution, Equation 11-4, we have to obey the condition, $L/R_o > 10$. We obtained the centerline temperature of the pine logs which vary in their radius from 0.2 m to 0.3 m and every other parameter staying the same as given above by using the one-term approximation in Equation 11-10. Time elapsed in hours to reach from 20 C

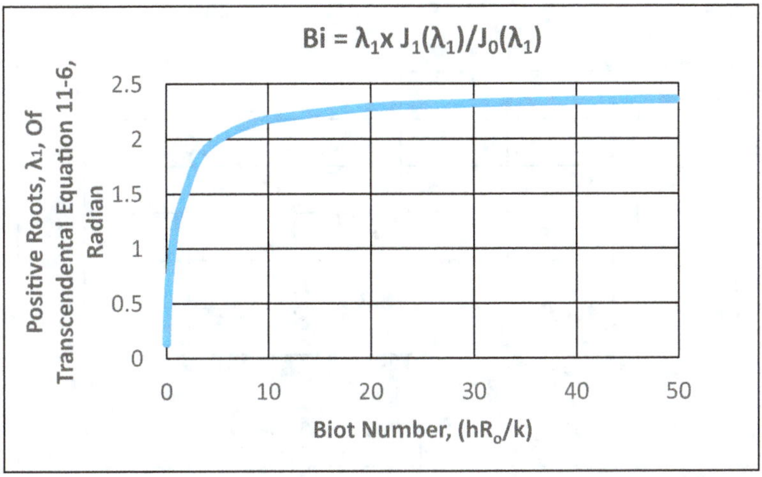

Figure 11-2 λ_1 versus Biot number

to 150 C in a 200 C radiation environment in vacuum versus pine log radius in meter is shown in Figure 11-4. The sensitivity of time to reach from 20 C to 150 C in a 200 C radiation environment in vacuum to pine log's radius is 5.23 hours increase in elapsed time per 0.01-meter increase in radius. Therefore, solutions to Equation 11-4 are very sensitive to changes in pine log's radius.

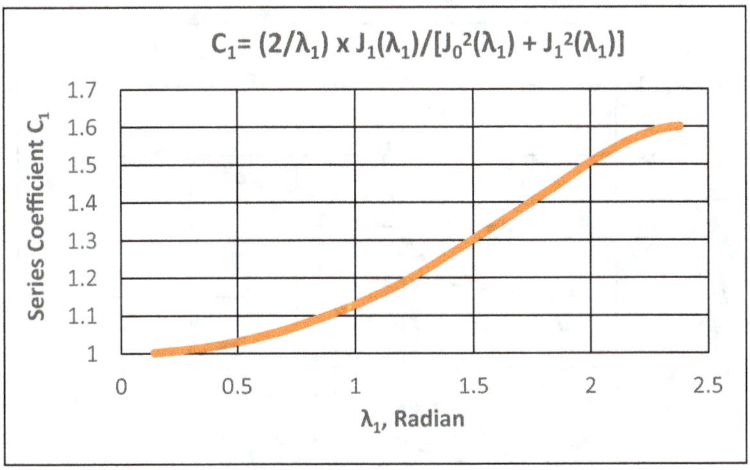

Figure 11-3 λ_1 versus C_1

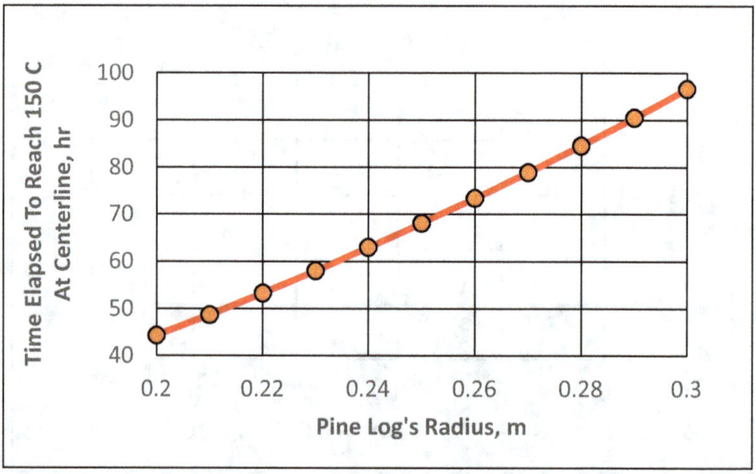

Figure 11-4 Time elapsed in hours to reach from 20 C to 150 C in a 200 C radiation environment in vacuum versus pine log radius in meter

The second variable we will investigate variations is the thermal conductivity, *k*, of pine log in the radial direction, namely perpendicular to the grain direction. *k* depends on the moisture content in the pine log and varies between 0.1 to 0.2 W/(m-K). We obtained the centerline temperature of the pine logs which vary in their *k* from 0.1 to 0.2 W/(m-K) and every other parameter stays the same

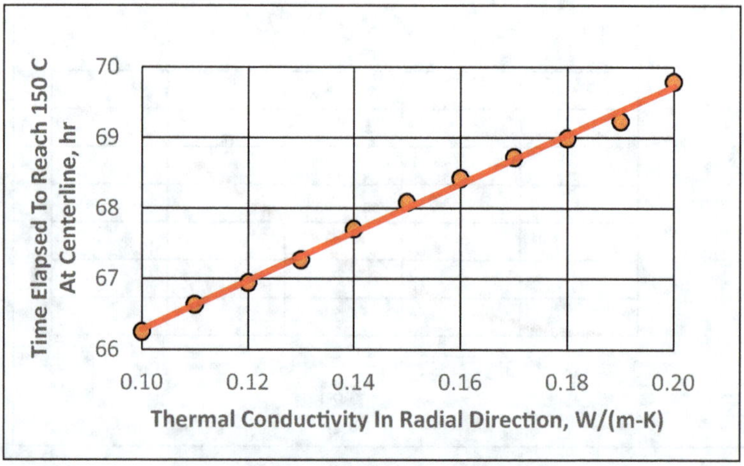

Figure 11-5 Time elapsed to reach 150 C at centerline of pine log versus thermal conductivity in the radial direction

as given above by using the one-term approximation in Equation 11-10. Time elapsed in hours to reach from 20 C to 150 C in a 200 C radiation environment in vacuum versus pine log's k is shown in Figure 11-5.

The sensitivity of time to reach from 20 C to 150 C in a 200 C radiation environment in vacuum to pine log's k is 20.65 minutes increase in elapsed time per 0.01-W/(m-K) increase in its radial thermal conductivity. Therefore, solutions to Equation 11-4 are not very sensitive to changes in pine log's radial thermal conductivity.

12

Roasting A Piece of Beef Which Has a Short Cylindrical Shape

In this chapter, we will investigate roasting a piece of lean beef which has a short cylindrical shape. Roasting is done in a convection oven at a temperature of 200 C. The chef wants to know the time it takes to roast one piece beef as rare and another same size piece of beef as well done. Dimensions of the short cylindrical beef piece in brown color is shown in Figure 12-1.

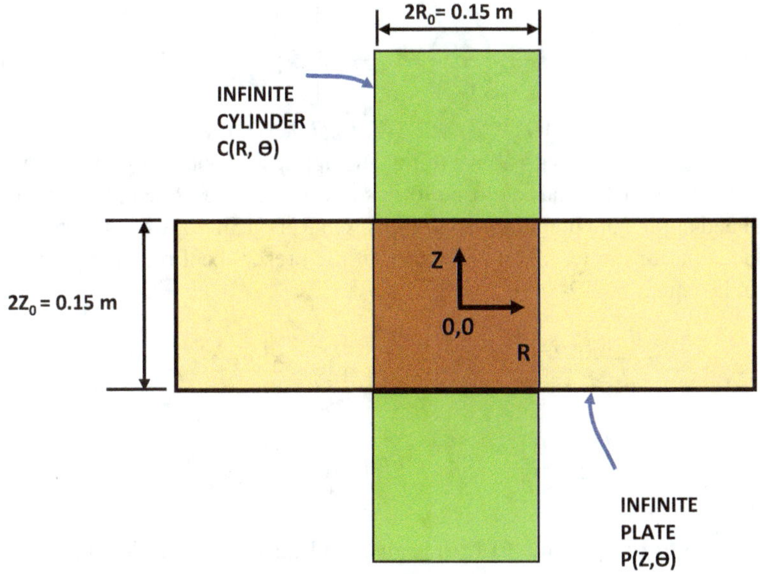

Figure 12-1 Short cylindrical shape of a beef piece in brown color to be roasted as the product of an infinite cylinder and an infinite plate. (0, 0) is the same center for the infinite cylinder in green color and for the infinite plate in yellow color.

Since beef pieces are initially at a uniform temperature, they also are uniformly heated all around and beef's radial and circumferential thermal conductivities are the same, no heat transfer occurs in the circumferential direction and beef's temperature is functions of radial direction, R, vertical direction, Z, and time, θ, as shown in Figure 12-1 and as represented in Equation 12-1. The cylindrical shape of the beef piece is the product of an infinite cylinder, C(R,θ), and an infinite plate, P(Z,θ), for exact temperature-time solutions.

$$\frac{1}{\alpha}\frac{\partial T}{\partial \theta}=\frac{\partial^2 T}{\partial R^2}+\frac{1}{R}\frac{\partial T}{\partial R}+\frac{\partial^2 T}{\partial Z^2} \qquad \text{12-1}$$

where $T(R, Z, \theta)$ is the temperature of the beef in centigrade and α is the constant thermal diffusivity of the beef in m²/s with initial and boundary conditions given in Equations 12-2.

$$T_{ini}\left(R,Z,0\right)=5C \qquad \text{12-2a}$$

$$\left(\frac{\partial T}{\partial R}\right)_{R=0}=0 \qquad \text{12-2b}$$

$$-k\left(\frac{\partial T}{\partial R}\right)_{R=R_o}=h_{rad}\left(T_\infty-T\right)_{R=R_0} \qquad \text{12-2c}$$

$$\left(\frac{\partial T}{\partial Z}\right)_{Z=0}=0 \qquad \text{12-2d}$$

$$-k\left(\frac{\partial T}{\partial Z}\right)_{Z=\pm L}=h\left(T_\infty-T\right)_{Z=\pm L} \qquad \text{12-2e}$$

$T(R, Z, \theta)$ is a product of $T(R, \theta)$ and $T(Z, \theta)$. $T(R, \theta)$ has a closed form solution for an infinite cylinder with above initial and applicable boundary conditions, see Ref [15]. Also, $T(Z, \theta)$ has a closed form solution for an infinite plate with above initial and applicable boundary conditions, see Ref. [15]. So, we can obtain exact product solutions to temperature history of the present beef roasting problem as shown in Equations 12-3.

$$\frac{T\left(R,Z,\theta\right)-T_{oven}}{T_{ini}-T_{oven}}=\left(\frac{T\left(R,\theta\right)-T_{oven}}{T_{ini}-T_{oven}}\right)_{\substack{infinite\\cylinder}}\times\left(\frac{T\left(Z,\theta\right)-T_{oven}}{T_{ini}-T_{oven}}\right)_{\substack{infinite\\plate}} \qquad \text{12-3a}$$

or

$$\frac{T\left(R,Z,\theta\right)-T_{oven}}{T_{ini}-T_{oven}}=C\left(R,\theta\right)_{\substack{infinite\\cylinder}}\times P\left(Z,\theta\right)_{\substack{infinite\\plate}} \qquad \text{12-3b}$$

For an **infinite cylinder** $\boldsymbol{C(R, \theta)}$ is given in Equations 12-4. See Ref. [10].

$$\frac{T\left(R,\theta\right)-T_{oven}}{T_{ini}-T_{oven}}=\sum_{i=1}^{\infty}A_i e^{-\lambda_i^2 FO_c} J_0\left(\lambda_i\frac{R}{R_0}\right) \qquad \text{12-4a}$$

$$A_i = \frac{2}{\lambda_i} \times \frac{J_1(\lambda_i)}{J_0^2(\lambda_i) + J_1^2(\lambda_i)} \quad \text{12-4b}$$

$$\lambda_i \frac{J_1(\lambda_i)}{J_0(\lambda_i)} = Biot\ Number_c \quad \text{12-4c}$$

where J_0 and J_1 are Bessel functions of the first kind. Since $J_0(0) = 1$, the centerline temperature of the infinite cylinder at $J_0(0)$ becomes as shown in Equation 12-5. The non-dimensional time, the Fourier number, for an infinite cylinder is $FO_c = \frac{a\theta}{R_o^2}$. The non-dimensional Biot number for an infinite cylinder, the ratio of conduction to convection resistances is, $Biot\ Number_c = \frac{hR_o}{k}$.

$$\frac{T(R=0,\theta) - T_{oven}}{T_{ini} - T_{oven}} = \sum_{i=1}^{\infty} A_i e^{-\lambda_i^2 FO_c} \quad \text{12-5}$$

For an **infinite plate** $\boldsymbol{P(Z, \theta)}$ is given in Equations 12-6. See Ref. [10].

$$\frac{T(Z,\theta) - T_{oven}}{T_{ini} - T_{oven}} = \sum_{i=1}^{\infty} E_i e^{-\gamma_i^2 FO_p} cos\left(\gamma_i \frac{Z}{Z_0}\right) \quad \text{12-6a}$$

$$E_i = \frac{4\,sin(\gamma_i)}{2\gamma_i + sin(2\gamma_i)} \quad \text{12-6b}$$

$$\gamma_i tan(\gamma_i) = Biot\ Number_p \quad \text{12-6c}$$

The centerline temperature for the infinite plate at cos(0) = 1 becomes as shown in Equation 12-7.

$$\frac{T(Z=0,\theta) - T_{oven}}{T_{ini} - T_{oven}} = \sum_{i=1}^{\infty} E_i e^{-\gamma_i^2 FO_p} \quad \text{12-7}$$

The non-dimensional time, the Fourier number, for an infinite plate is $FO_p = \frac{a\theta}{Z_o^2}$. The non-dimensional Biot number for an infinite plate, the ratio of conduction to convection resistances, is $Biot\ Number_p = \frac{hZ_o}{k}$.

Eigenvalues for λ_1 and γ_i are positive roots of transcendental Equations 12-4c and 12-6c, respectively. Times for roasting the beef as rare and as well done and beef's temperature distributions are determined by using first term approximations of Equations 12-5 and 12-7 and by using the following constant beef properties.

Beef radius: $R_0 = 0.075\,m$; Beef length: $Z_0 = 0.075\,m$
Initial beef temperature: $T(R,Z,0)_{ini} = 5C$; Oven temperature: $T_{oven} = 200C$

Convection heat transfer coefficient: $h = 20W/(m^2 - C)$
Beef's thermal conductivity: $k = 0.43W/(m - C)$
Beef's thermal diffusivity: $\alpha = 6.71 \times 10^{-7}\, m^2/s$
Chef requires that rare roasting of beef is completed when $T(0,0,\theta_{rare}) = 80C$.
Chef also requires that well done roasting of beef is completed when $T(0,0,\theta_{well\ done}) = 100C$.
In order to be able to use first term center temperature approximations for $\frac{T(R=0,\theta)-T_{oven}}{T_{ini}-T_{oven}}$ and for $\frac{T(Z=0,\theta)-T_{oven}}{T_{ini}-T_{oven}}$, we have to make sure that their respective Fourier numbers are greater than 0.2, namely $FO_c > 0.2$ and $FO_p > 0.2$.

$$\theta_c > \frac{0.2 \times 0.075^2}{6.71 \times 10^{-7}} = 1{,}678\, s \quad \text{and} \quad \theta_p > \frac{0.2 \times 0.075^2}{6.71 \times 10^{-7}} = 1{,}678\, s$$

First, let us determine necessary values in the first term for infinite cylinder, namely, $C(0,\theta) = \frac{T(R=0,\theta)-T_{oven}}{T_{ini}-T_{oven}} = A_1 e^{-\lambda_1^2 FO_c}$ from Equation 12-5. Charts to obtain λ_1 and A_1, used values of Bessel functions of the first kind, J_0 and J_1, given in Figure 12-2.
λ_1 = 1.847 in Equation 12-5 is obtained from the chart given in Figure 12-3 for a Biot number of $Biot\ Number_c = \frac{hR_o}{k} = \frac{20 \times 0.075}{0.43} = 3.488$. Then A_1 = 1.444 is obtained from the chart given in Figure 12-4 for $\lambda_1 = 1.847$.

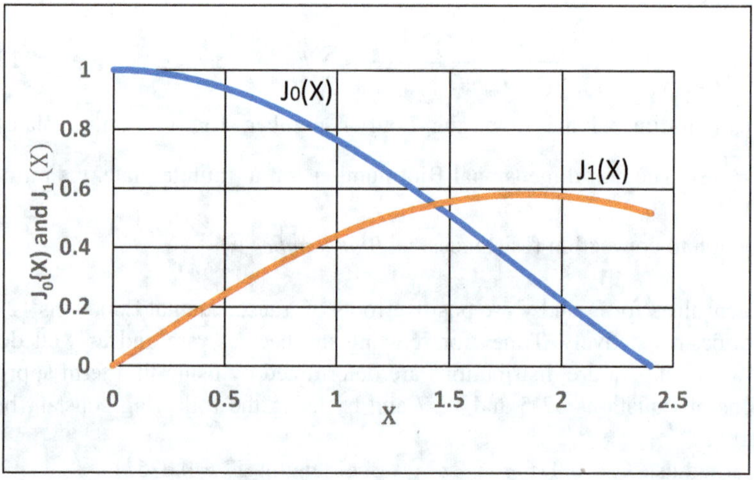

Figure 12-2 Positive values of Bessel functions of the first kind, J_0 (blue line) and J_1 (orange line)

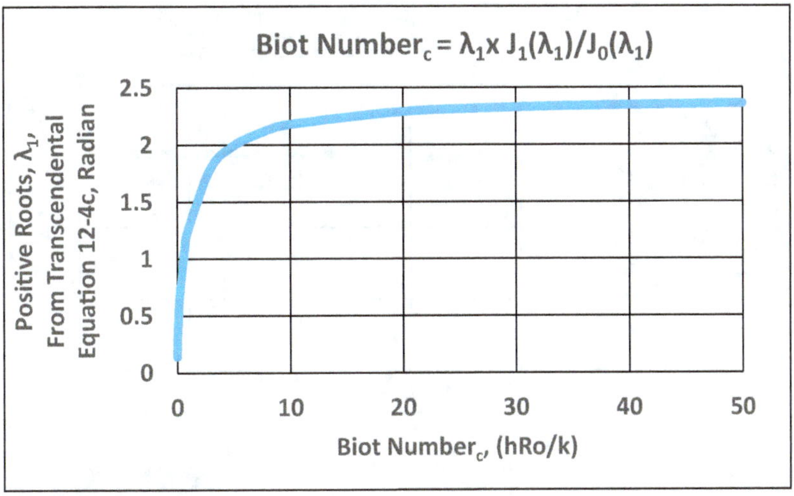

Figure 12-3 Eigenvalue λ_1 as a function of Biot number$_c$

The first term for infinite cylinder becomes as shown in Equation 12-8.

$$C(0,\theta)=\frac{T(R=0,\theta)-T_{oven}}{T_{ini}-T_{oven}}=1.444e^{-1.847^2 FO_c} \qquad 12\text{-}8$$

Second, let us determine necessary values in the first term for the infinite plane, namely, $P(0,\theta)=\dfrac{T(Z=0,\theta)-T_{oven}}{T_{ini}-T_{oven}}=E_1e^{-\gamma_1^2 FO_p}$ from Equation 12-7.

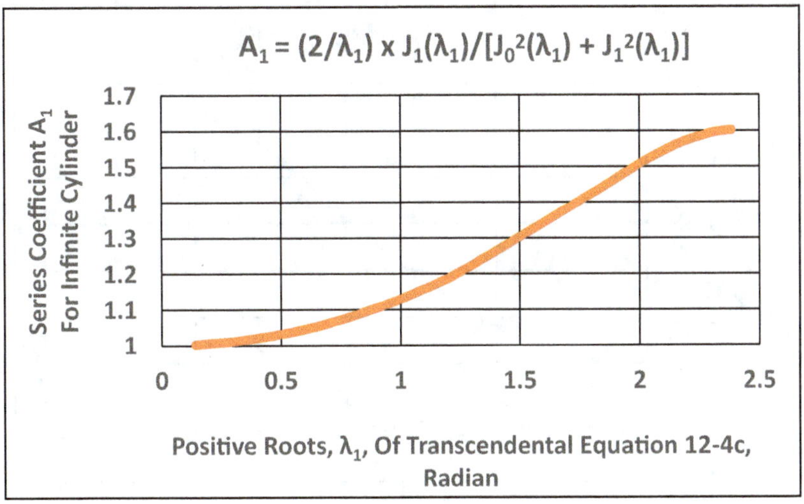

Figure 12-4 Series constant A_1 in Equation 12-4b as a function of λ_1

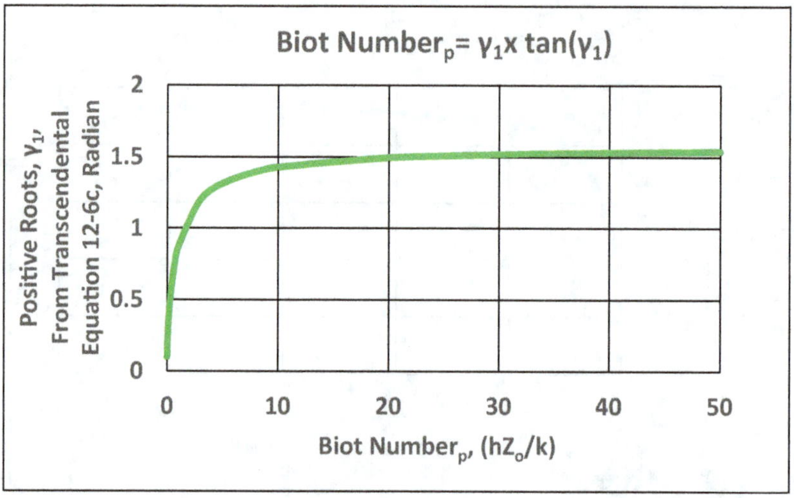

Figure 12-5 Eigenvalue γ_1 as a function of Biot number

γ_1 = 1.228 in Equation 12-7 is obtained from the chart given in Figure 12-5 for a Biot number of $Biot\ Number_p = \frac{hZ_o}{k} = \frac{20 \times 0.075}{0.43} = 3.488$. Then $E_1 = 1.219$ is obtained from the chart given in Figure 12-6 for $\gamma_1 = 1.228$.

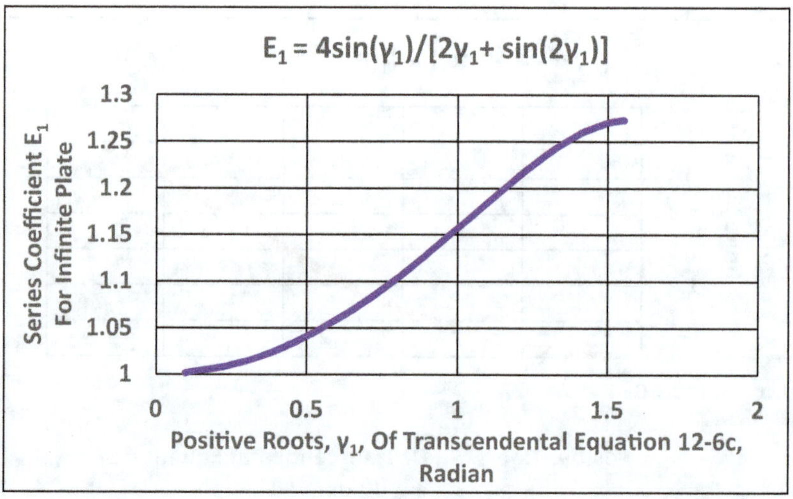

Figure 12-6 Series constant E_1 in Equation 12-6b as a function of γ_1

The first term for infinite plate becomes as shown in Equation 12-9.

$$P(0,\theta)=\frac{T(Z=0,\theta)-T_{oven}}{T_{ini}-T_{oven}}=1.219e^{-1.228^2 FO_p} \quad 12\text{-}9$$

From the product of Equations 12-8 and 12-9, we can determine times that the center of the short cylindrical shaped beef reach 80 C and 100 C in a 200 C convection oven as shown in Equations 12-10.

$$\frac{T(0,0,\theta)-200}{5-200}=C(0,\theta)_{\substack{infinite\\cylinder}}\times P(0,\theta)_{\substack{infinite\\plate}} \quad 12\text{-}10a$$

$$\frac{T(0,0,\theta)-200}{5-200}=1.444e^{-1.1847^2 FO_c}\times 1.219e^{-1.228^2 FO_p} \quad 12\text{-}10b$$

Non-dimensional center temperatures for $C(0,\theta)_{\substack{infinite\\cylinder}}$, for $P(0,\theta)_{\substack{infinite\\plate}}$ and for $C(0,\theta)_{\substack{infinite\\cylinder}}\times P(0,\theta)_{\substack{infinite\\plate}}$ versus time are shown in Figure 12-7.

For rare roasting, the non-dimensional center temperature of the beef piece is $\frac{80-200}{5-200}=0.6154$. It takes 1792.6 s (29.9 min) for the center of the beef piece to reach 80 C as shown with green dot in Figure 12-7. For well-done roasting, the non-dimensional center temperature of the beef piece is $\frac{100-200}{5-200}=0.5128$. It takes 2103.5 s (35.1 min) for the center of the beef piece to reach 100 C as shown with red dot in Figure 12-7.

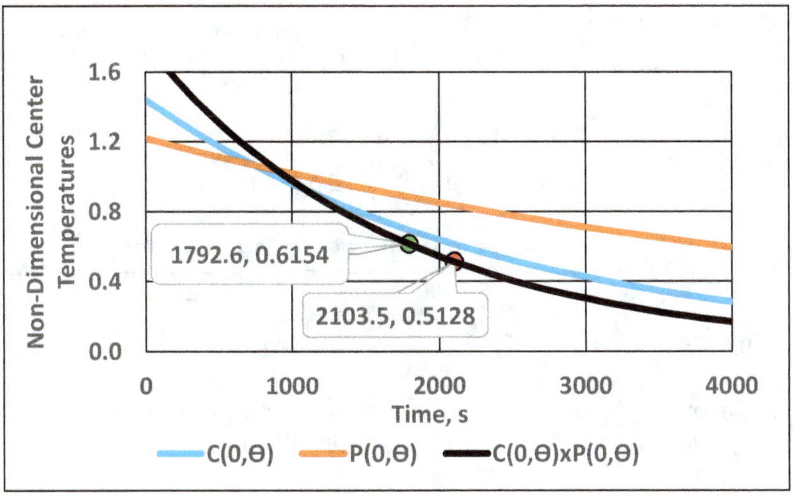

Figure 12-7 Non-dimensional center temperature for infinite cylinder, for infinite plate and for product of infinite cylinder and infinite plate versus time

Above one-term approximations are valid for C(0, Θ), Equation 12-8, and for P(0,Θ), Equation 12-9, since center temperature times to reach 80 C and 100 C are greater than 1,678 seconds or $FO_c > 0.2$ and $FO_p > 0.2$.

Now, let us determine radial temperature distributions in the rarely roasted beef at three locations, namely at *T*(*R*,0,1792.6), *T*(*R*,0.0375,1792.6), and *T*(*R*,0.075,1792.6).

The normalized radial temperature distribution at $Z = 0$ is obtained from Equation 12-3b and it is given in Equation 12-11.

$$\frac{T(R,0,1792.6)-200}{5-200} = C(R,1792.6)_{\substack{infinite\\cylinder}} \times P(0,1792.6)_{\substack{infinite\\plate}} \qquad \text{12-11}$$

Using the first term approximation for the infinite cylinder from Equation 12-4a, $C(R,1792.6)_{\substack{infinite\\cylinder}}$ becomes as shown in Equation 12-12.

$$\frac{T(R,1792.6)-200}{5-200} = C(0,1792.6)_{\substack{infinite\\cylinder}} \times J_0\left[1.847\left(\frac{R}{0.075}\right)\right] \qquad \text{12-12}$$

where $C(0,1792.6)_{\substack{infinite\\cylinder}} = 0.6965$. Also, $P(0,1792.6)_{\substack{infinite\\plate}} = 0.8835$. Using these two constant centerline temperature values and Equation 12-12, we can now solve Equation 12-11 for the radial temperature distribution at $Z = 0$ when *T*(0, 0, 1792.6) = 80 *C* as shown in red-color in Figure 12-8.

Next, let us determine radial temperature distributions in the rare roasted beef at Z = 0.0375 *m*, namely at *T*(*R*, 0.0375,1792.6). First, we have to determine the temperature of the beef piece at Z = 0.0375 *m* for the infinite plate using the first term approximation from Equation 12-6a for infinite plate as shown in Equation 12-13.

$$P(0.0375,1792.6) = \frac{T(0.0375,1792.6)-200}{T(0,1792.6)-200} = \cos\left(\gamma_1 \frac{0.0375}{0.075}\right) \qquad \text{12-13}$$

Equation 12-13 is evaluated as follows.

$$\cos\left(1.228\frac{0.0375}{0.075}\right) = 0.8174$$

$$P(0.0375,1792.6) = \frac{T(0.0375,1792.6)-200}{5-200} = \frac{T(0,1792.6)-200)}{(5-200)} \times 0.8174$$

$$P(0.0375,1792.6) = \frac{T(0.0375,1792.6)-200)}{5-200} = 0.8835 \times 0.8174 = 0.7222$$

The normalized radial temperature distribution at $Z = 0.0375$ *m* is obtained from Equation 12-3b and it is given in Equations 12-14.

$$\frac{T(R,0.0375,1792.6)-200}{5-200} = C(R,1792.6)_{\substack{infinite\\cylinder}} \times P(0.0375,1792.6)_{\substack{infinite\\plate}} \qquad \text{12-14a}$$

or

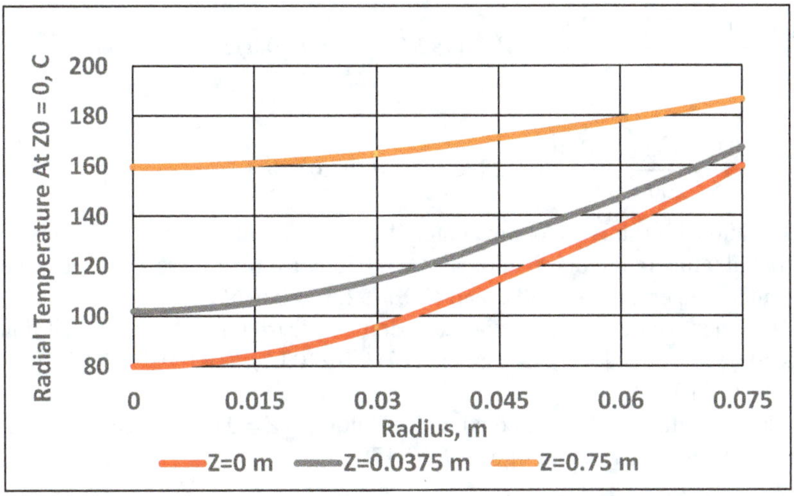

Figure 12-8 Radial temperature distributions for rare roasted beef when its center temperature is at 80 C.

$$\frac{T(R,0.0375,1792.6)-200}{5-200}=0.7222\times C(0,1792.6)\times J_0\left(1.847\frac{R}{0.075}\right) \quad \text{12-14b}$$

Equation 12-14b is the gray-colored curve in Figure 12-8 which represents the radial temperature distribution at $Z = 0.0375\ m$ for the rare roasted beef when its center temperature is at 80 C and $C(0,1792.6) = 0.6965$.

Next, let us determine radial temperature distributions in the rarely roasted beef at $Z = 0.075\ m$, namely at $T(R,0.075,1792.6)$. First, we have to determine the temperature of the beef piece at $Z = 0.075\ m$ for the infinite plate. Using the first term approximation from Equation 12-6a for infinite plate, we get Equation 12-15.

$$P(0.075,1792.6)=\frac{T(0.075,1792.6)-200}{T(0,1792.6)-200}=\cos\left(\gamma_1\frac{0.075}{0.075}\right) \quad \text{12-15}$$

Equation 12-15 is evaluated as follows.

$$\cos\left(1.228\frac{0.075}{0.075}\right)=0.3364$$

$$P(0.075,1792.6)=\frac{T(0.075,1792.6)-200)}{(5-200)}=\frac{T(0,1792.6)-200)}{(5-200)}\times 0.3364$$

$$P(0.075,1792.6)=\frac{T(0.075,1792.6)-200)}{(5-200)}=0.8835\times 0.3364=0.2972$$

The normalized radial temperature distribution at $Z = 0.075\ m$ is obtained from Equation 12-3b and it is given in Equations 12-16.

$$\frac{T(R,0.075,1792.6)-200}{5-200} = C(R,1792.6)_{\substack{infinite\\cylinder}} \times P(0.075,1792.6)_{\substack{infinite\\plate}} \quad \text{12-16a}$$

or

$$\frac{T(R,0.075,1792.6)-200}{5-200} = 0.2972 \times C(0,1792.6) \times J_0\left(1.847\frac{R}{0.075}\right) \quad \text{12-16b}$$

Equation 12-16b is the orange-colored curve in Figure 12-8 which represents the radial temperature distribution at $Z = 0.075$ *m* for the rare roasted beef when its center temperature is at 80 C and $C(0,1792.6) = 0.6965$.

Now, let us determine radial temperature distributions in the well-done roasted beef at three locations, namely at $T(R,0,2103.5)$, $T(R,0.0375,2103.5)$, and $T(R,0.075,2103.5)$.

The normalized radial temperature distribution at $Z = 0$ is obtained from Equation 12-3b and it is given in Equation 12-17.

$$\frac{T(R,0,2103.5)-200}{5-200} = C(R,2103.5)_{\substack{infinite\\cylinder}} \times P(0,2103.5)_{\substack{infinite\\plate}} \quad \text{12-17}$$

Using the first term approximation for the infinite cylinder from Equation 12-4a, $C(R,2103.5)_{\substack{infinite\\cylinder}}$ becomes as shown in Equation 12-18.

$$\frac{T(R,2103.5)-200}{5-200} = C(0,2103.5)_{\substack{infinite\\cylinder}} \times J_0\left[1.847\left(\frac{R}{0.075}\right)\right] \quad \text{12-18}$$

where $C(0,2103.5)_{\substack{infinite\\cylinder}} = 0.6138$. Also, $P(0,2103.5)_{\substack{infinite\\plate}} = 0.8355$.

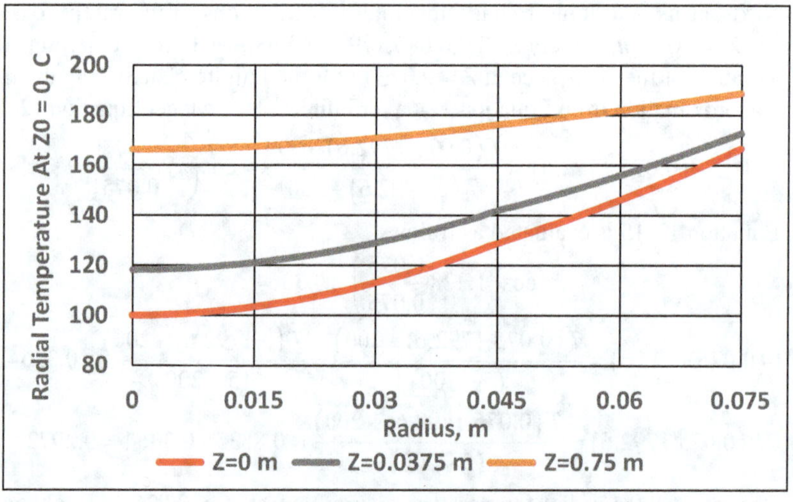

Figure 12-9 Radial temperature distributions for well-done roasted beef when its center temperature is at 100 C.

Using these two constant centerline temperature values and Equation 12-18, we can now solve Equation 12-17 for the radial temperature distribution at $Z=0$ when $T(0,0,2103.5)=100\ C$ as shown in red-color in Figure 12-9. For well-done roasting, the non-dimensional center temperature of the beef piece is $\frac{100-200}{5-200}=0.5128$. It takes 2103.5 s (35.1 min) for the center of the beef piece to reach 100 C as shown with red dot in Figure 12-7. We can again use on-term approximations for Equations 12-4a and 12-6a since $\Theta > 1{,}678\ s$.

Next, let us determine radial temperature distributions in the well-done roasted beef at $Z = 0.075\ m$, namely at $T(R,0{,}0375{,}2103.5)$. First, we have to determine the temperature of the beef piece at $Z = 0.075\ m$ for infinite plate. Using the first term approximation from Equation 12-6a for infinite plate, we get Equation 12-19.

$$P(0.0375,2103.5)=\frac{T(0.0375,2103.5)-200}{T(0,2103.5)-200}=\cos\left(\gamma_1\frac{0.0375}{0.075}\right) \qquad 12\text{-}19$$

Equation 12-19 is evaluated as follows.

$$\cos\left(1.228\frac{0.0375}{0.075}\right)=0.8174$$

$$P(0.0375,2103.5)=\frac{T(0.0375,2103.5)-200}{5-200}=\frac{T(0,2103.5)-200}{5-200}\times 0.8174$$

$$P(0.0375,2103.5)=\frac{T(0.0375,2103.5)-200}{5-200}=0.8355\times 0.8174=0.6830$$

The normalized radial temperature distribution at $Z = 0.075\ m$ is obtained from Equation 12-3b and it is given in Equations 12-20.

$$\frac{T(R,0.0375,2103.5)-200}{5-200}=C(R,2103.5)_{\substack{infinite\\cylinder}}\times P(0.0375,2103.5)_{\substack{infinite\\plate}} \qquad 12\text{-}20a$$

or

$$\frac{T(R,0.0375,2103.5)-200}{5-200}=0.6830\times C(0,2103.5)\times J_0\left(1.847\frac{R}{0.075}\right) \qquad 12\text{-}20b$$

Equation 12-20b is the gray-colored curve in Figure 12-9 which represents the radial temperature distribution at $Z = 0.075\ m$ for the well-done roasted beef when its center temperature is at 100 C, and $C(0,2103.5) = 0.6138$.

Next, let us determine radial temperature distributions in the well-done roasted beef at $Z=0.075\ m$, namely at $T(R, 0.075{,}2103.5)$. First, we have to determine the temperature of the beef piece at $Z = 0.075\ m$ for the infinite plate. Using the first term approximation from Equation 12-6a for infinite plate, we get Equation 12-21.

$$P(0.075,2103.5)=\frac{T(0.075,2103.5)-200}{T(0,2103.5)-200}=\cos\left(\gamma_1\frac{0.075}{0.075}\right) \qquad 12\text{-}21$$

Equation 12-21 is evaluated as follows.

$$\cos\left(1.228\frac{0.075}{0.075}\right) = 0.3364$$

$$P(0.075, 2103.5) = \frac{T(0.075, 2103.5) - 200}{(5-200)} = \frac{T(0, 2103.5) - 200}{(5-200)} \times 0.3364$$

$$P(0.075, 2103.5) = \frac{T(0.075, 2103.5) - 200}{(5-200)} = 0.8355 \times 0.3364 = 0.2811$$

The normalized radial temperature distribution at Z = 0.075 m is obtained from Equation 12-3b and it is given in Equations 12-22.

$$\frac{T(R, 0.075, 2103.5) - 200}{5-200} = C(R, 2103.5)_{\substack{infinite \\ cylinder}} \times P(0.075, 2103.5)_{\substack{infinite \\ plate}} \qquad \text{12-22a}$$

or

$$\frac{T(R, 0.075, 2103.5) - 200}{5-200} = 0.2811 \times C(0, 2103.5) \times J_0\left(1.847\frac{R}{0.075}\right) \qquad \text{12-22b}$$

Equation 12-22b is the orange-colored curve in Figure 12-9 which represents the radial temperature distribution at Z = 0.075 m for the well-done roasted beef when its center temperature is at 100 C and $C(0,2103.5)=0.6138$.

Transient temperature distributions in a short cylindrical beef are well represented by $C(R, \Theta) \times P(Z, \Theta)$ in the above case since heat transfer is significant both in radial and in vertical directions. We were able to use one-term approximations to Equation 12-4a for infinite cylinder and Equation 12-6a for infinite plate since both $FO_c > 0.2$ and $FO_p > 0.2$.

13

Frost Damage in a Spherical Orange

In this chapter, we will investigate how frost damage penetrates a spherical and ripe orange. We assume that there is a sudden change in outside temperature at night with convection heat transfer at the surface of the orange. Also, there is a cloudy sky so there is minimal radiation heat loss to the night sky. The temperature change in the spherical orange occurs only in the radial direction. So, orange's temperature is only functions of radius, R, and time, θ, as given in Equation 13-1 in spherical coordinates. Inside the orange, there are no temperature gradients in polar and azimuthal directions.

$$\frac{\partial^2 T}{\partial R^2}+\frac{2}{R}\frac{\partial T}{\partial R}=\frac{1}{\alpha}\frac{\partial T}{\partial \theta} \qquad 13\text{-}1$$

where $T(R, \theta)$ is the temperature of the spherical orange in centigrade and α is the constant thermal diffusivity of orange's flesh and skin in m^2/s with initial and boundary conditions given in Equations 13-2.

$$T_{ini}(R,0)=15C \qquad 13\text{-}2a$$

$$\left(\frac{\partial T}{\partial R}\right)_{R=0}=0 \qquad 13\text{-}2b$$

$$-k\left(\frac{\partial T}{\partial R}\right)_{R=R_o}=h\left(T_{env}-T\right)_{R=R_0} \qquad 13\text{-}2c$$

where k is the constant thermal conductivity of orange's flesh and skin in W/(m-C) and h is the average convection heat transfer at the surface of the orange in W/(m^2-C). For a sphere $T(R, \theta)$ that satisfy Equations13-1 and 13-2, is given in Equations 13-3. See Ref. [10].

$$\frac{T(R,\theta)-T_{env}}{T_{ini}-T_{env}}=\sum_{i=1}^{\infty}A_i e^{-\lambda_i^2 FO}\left[\frac{\sin\left(\lambda_i \dfrac{R}{R0}\right)}{\left(\lambda_i \dfrac{R}{R0}\right)}\right] \qquad 13\text{-}3a$$

$$A_i = \frac{4\left[sin(\lambda_i) - \lambda_i cos(\lambda_i)\right]}{2\lambda_i - sin(2\lambda_i)} \quad \text{13-3b}$$

$$1 - \lambda_i cot(\lambda_i) = Biot\ Number \quad \text{13-3c}$$

The center temperature for the orange is given in Equation 13-4 when $limit_{R\to 0} \frac{\sin\left(\lambda_i \frac{R}{R0}\right)}{\left(\lambda_i \frac{R}{R0}\right)} = \frac{0}{0} = 1$. The non-dimensional time, the Fourier number, for a sphere is $FO = \frac{a\theta}{R_0^2}$. The non-dimensional Biot number for a sphere, the ratio of conduction to convection resistances is $\frac{hR_o}{k}$.

$$\frac{T(R=0,\theta) - T_{env}}{T_{ini} - T_{env}} = \sum_{i=1}^{\infty} A_i e^{-\lambda_i^2 FO} \quad \text{13-4}$$

The spherical orange has a diameter of 0.08 m or $R_0 = 0.04\ m$ and degree of frost penetration is shown in Figure 13-1 according to USA grade specifications. See Ref. [17].

Step 1: Cut a 2 mm slice of the stem end to take the skin off and expose the flesh.

Step 2: Cut a ¼ inch thick slice (6.35 mm thick) off the exposed flesh. If frost damage, dryness in fruit's segments, is not seen on the freshly cut surface, then

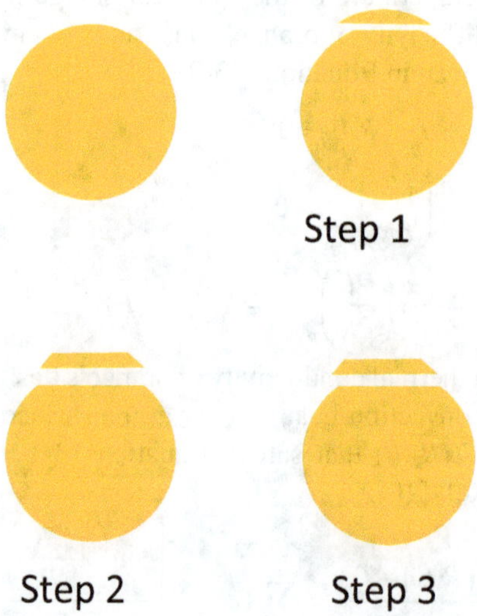

Figure 13-1 USA frost damage specifications for oranges

this is a grade 1 orange. If frost damage is seen on this surface, then go deeper into the flesh to Step 3.

Step 3: Cut another ¼ inch thick slice (6.35 mm thick). If frost damage, dryness in fruit's segments, is not seen on the freshly cut surface, then this is a grade 2 orange.

There is a frost damage risk, if orange flesh experiences temperatures below –2 C.

First, we will investigate the drop in center temperature of an orange at with a 0.04 m radius.

Second, we will investigate when flesh of orange goes below –2 C at the radius of 0.03165 m or after a ¼ inch thick slice (0.00635 m thick) is cut off after flesh is exposed, namely Step 2 above for grade 1 frost inspection.

Third, we will investigate when flesh of orange goes below –2 C at radius of 0.0253 m or after another ¼ inch thick slice (0.00635 m thick) is cut off after the exposed flesh, namely Step 3 above for grade 2 frost inspection.

We will use following constant properties during our calculations for fruit's different segments and for convection heat transfer at skin's surface.

$$Ro = 0.04\,m;\;\; k = 0.57\,W/(m-C);\;\; \alpha = 1.39\times10^{-7}\,m^2/s$$

$$h = 15\ W/(m^2 - \mathrm{C})$$

The Biot number for this investigation becomes as given in Equation 13-5.

$$Biot\ Number = \frac{hR_0}{k} = 1.053 \qquad 13\text{-}5$$

Since *Biot Number* > 0.1, we cannot use the lumped capacitance method to solve the present spherical orange problem. We have to consider the variation of temperature with respect to radius as given in Equation 13-1.

All eigen values, λ_i, and constants, A_i, that are defined in Equations 13-3 depend on the value of Biot number and they are provided in Ref. [15]. First five values of λ_i and A_i for spherical solutions given in Equations 13-3 are provided in Table 13-1.

Biot Number = 1.053		
i	λ_i	A_i
1	1.5988	1.2840
2	4.7233	–0.4442
3	7.8606	0.2675
4	11.0004	–0.1913
5	14.1409	0.1489

Table 13-1 Eigen values, λ_i, and constants, A_i, that are used in the present calculations

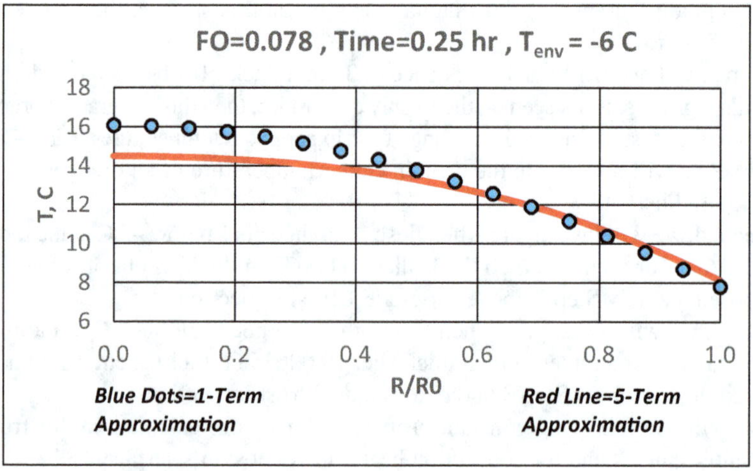

Figure 13-2 Radial temperature profile in a spherical orange after 900 seconds exposure to -6 C environment. One-term versus five-term approximations.

First term approximation of the series solution given in Equation 13-3a is valid, if $FO > 0.2$ or > 2,296 s. Otherwise, we can use up to five term approximation of Equations 13-3a and 13-4 with λ_i and A_i values provided in Table 13-1.

Figure 13-2 shows that one-term approximation of Equation 13-3a overestimates the temperature towards the middle of the orange. However, if we increase the exposure time to 1,800 s, overestimation of temperature towards

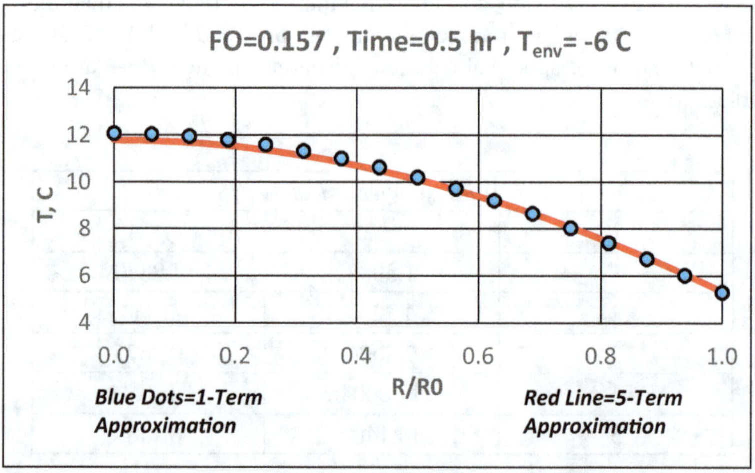

Figure 13-3 Radial temperature profile in a spherical orange after 1,800 seconds exposure to -6 C environment. One-term versus five-term approximations

the middle of the orange decreases and one-term approximation approaches values of five-term approximation as we get closer to $FO > 0.2$ or $\theta > 2{,}296$ s as shown in Figure 13-3.

The center temperature for an 8-cm diameter spherical orange is given in Figure 13-4 for different environment temperatures using Equation 13-4. When orange's center goes below -2 C, whole orange is frost damaged. With an initial temperature at 15 C, this orange becomes totally frost damaged after 3.14 hours, if the environment temperature suddenly drops to -4 C. It is totally frost damaged after 2.38 hours, if the environment temperature suddenly drops to -6 C, or totally frost damaged after 1.99 hours, if the environment temperature suddenly drops to -8 C, or totally frost damaged after 1.73 hours, if the environment temperature suddenly drops to -10 C.

Radial temperature distributions in the 8-cm diameter orange at different times when $T_{ini} = 15\ C$ and $T_{env} = -6\ C$ are shown in Figure 13-5. Up to 1.5 hours, orange segments do not experience any frost damage. At 2-hour mark, outer segments of the orange from $R = 3.25$ to $4.00\ cm$ fall below – 2 C frost damage temperature level and might experience frost damage at its outer segments, the orange-colored line. At 3-hour mark and afterwards, all segments of the orange fall below – 2 C frost damage temperature level and might experience frost damage throughout the whole orange, the green and black-colored lines.

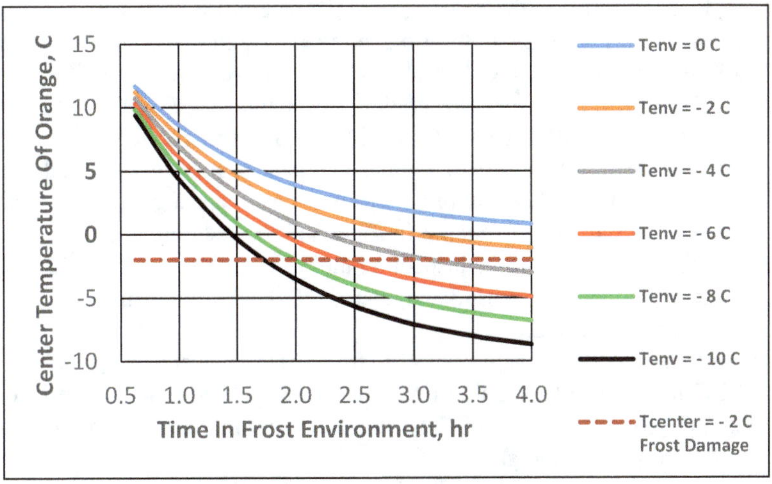

Figure 13-4 8-cm diameter orange's center temperature versus time profiles for different environment temperatures. -2C red dashed line represents the total frost damage line. When the center temperature of the orange goes below the -2 C line, total frost damage can occur.

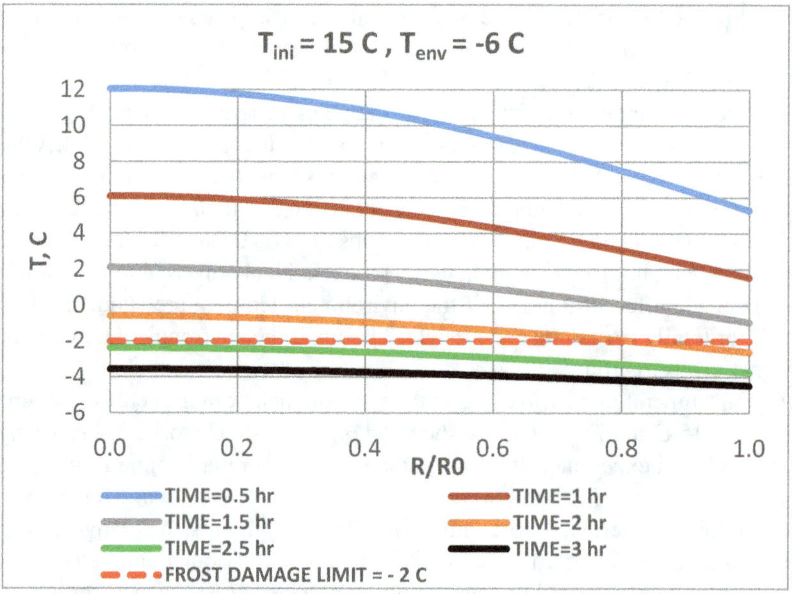

Figure 13-5 Radial temperature distributions in the 8-cm diameter orange at different times when T_{ini} = 15 C and T_{env} = -6 C

Next, let us examine radii limits for frost damage in USA orange grade specifications given in Figure 13-1. After a 2-mm skin is removed, Step 1 in Figure 13-1, a ¼ inch thick slice (0.00635 m thick) is removed off the exposed flesh, Step 2 in Figure 13-1, to inspect for grade 1 frost damage at $R = 0.03165$ m.

If there is no frost damage at $R = 0.03165$ m layer, then the orange can be accepted as grade 1. If there is frost damage in orange segments between the radii 0.03165 m to 0.038 m, then the orange cannot be accepted as grade 1 orange and the inspector has to go to Step 3 in Figure 13-1 by slicing another 1/4-inch (0.00635 m) thick layer, namely inspecting the fresh surface at $R = 0.0253$ m for frost damage. If there is no frost damage at $R = 0.0253$ m layer, then the orange can be accepted as grade 2.

Times that an orange reaches the -2 C frost damage level at radii 0.03165 m and 0.0253 m are calculated using Equations 13-3 for environment temperature of -3 C, -4 C, -6 C, -8 C and -10 C. The results are summarized in Table 13-2 and compared to results at in Figure 13-4.

Frost damage propagates from $R = 0.03165$ m to $R = 0.0253$ m in 0.13 hour and from $R = 0.0253$ m to orange's center in 0.22 hour by conduction heat transfer for all environment temperatures given above. The sensitivity of time to reach -2 C to environment's temperature for all three cases are the same as shown in Figure 13-6, since time dependence of temperature at any radius within the orange is the same as that of the center temperature in Equation 13-3a. The change in time

	R=0.03165 m Grade 1	R=0.0253 m Grade 2	R=0 m Whole Orange
Tenv = -3 C	3.57	3.70	3.92
Tenv = -4 C	2.77	2.90	3.12
Tenv = -6 C	2.03	2.16	2.38
Tenv = -8 C	1.64	1.77	1.99
Tenv = -10 C	1.38	1.51	1.73

Table 13.2 Times in hours that an orange with an 8-cm diameter falls below -2 C and susceptible to frost damage at radii 0.03165 m, 0.0253 m and 0 m for environment temperatures -3 C, -4 C, -6 C, -8 C and -10 C with an initial orange temperature of 15 C

to reach -2 C to change in environment's temperature increases fourfold as the environment's temperature increases from -10 C to -4 C.

The sensitivity of time to reach -2 C at the center of the orange to changes in heat transfer coefficient, h, is also investigated when $Tenv$ = -6 C. First term approximation of the series solution given in Equation 13-4 is used for these sensitivity calculations and $FO > 0.2$ or $\theta > 2{,}296$ s requirement is satisfied. As the heat transfer coefficient, h, changes, so does the Biot number. Eigen values, λ_i, and constants, A_i, are obtained from Figures 13-7 and 13-8 provided below, once the Biot number is known. See Ref. [15].

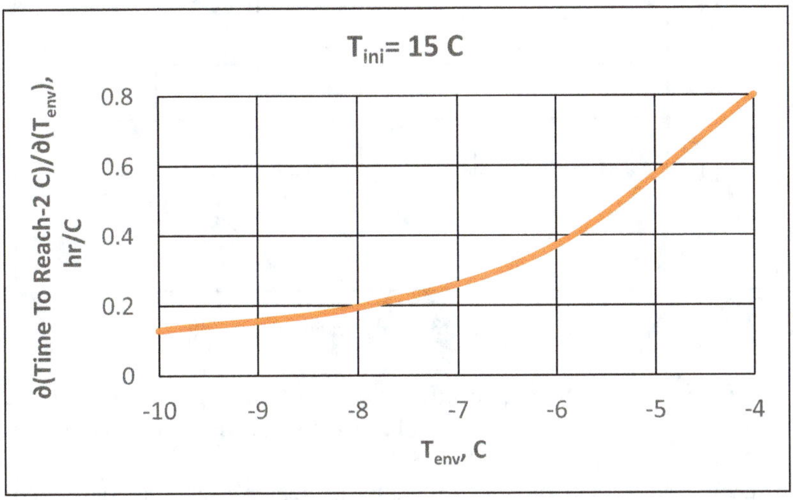

Figure 13-6 Sensitivity of time to reach -2 C to environment temperature for all three cases given in Table 13-2

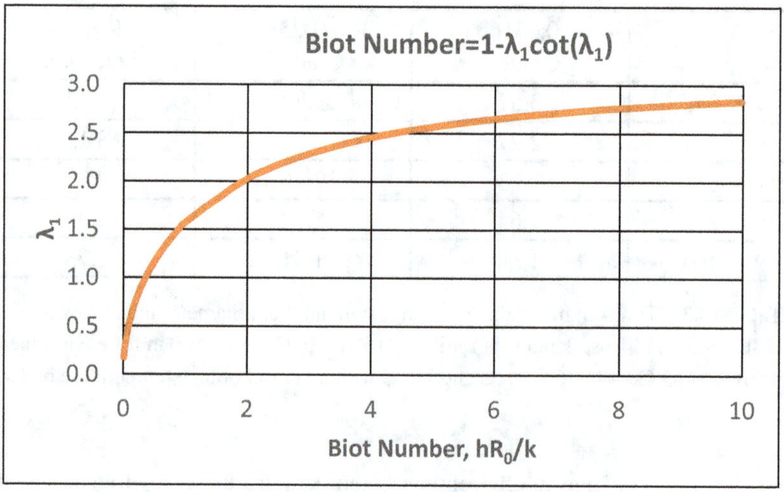

Figure 13-7 λ_i versus Biot number

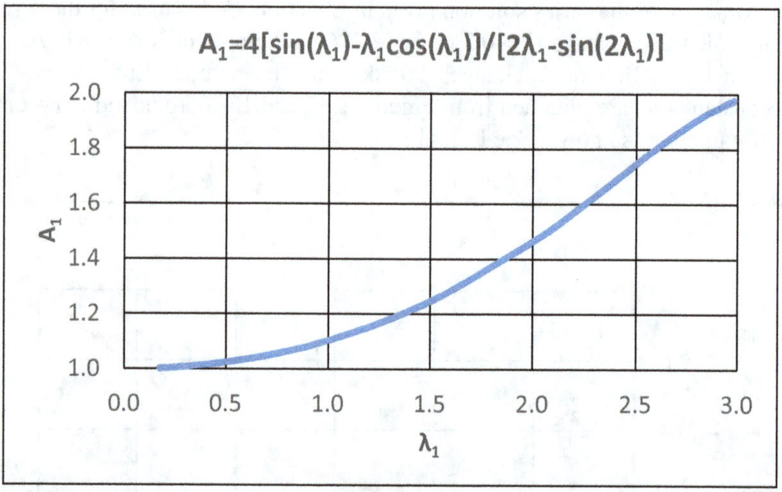

Figure 13-8 A_1 versus λ_i

The following heat transfer coefficients, *h*, are investigated for the present sensitivity analysis. Eigen values, λ_i, and constants, A_1, are given as a function of heat transfer coefficients, *h*, in Table 13-3.

Time to reach -2 C at the center of orange decreases as the heat transfer coefficient, *h*, at orange's surface increases as shown in Figure 13-9 when T_{ini} = 15 *C* and T_{env} = -6 *C*.

h, $W/(m^2\text{–}C)$	*Biot #*	λ_i	A_1
5	0.351	1.020	1.124
7.5	0.526	1.248	1.185
10	0.702	1.409	1.235
15	1.053	1.599	1.284
20	1.404	1.799	1.376
25	1.754	1.925	1.428
30	2.105	2.027	1.473
35	2.456	2.114	1.513
40	2.807	2.189	1.549
45	3.158	2.255	1.581
50	3.509	2.314	1.610

Table 13-3 Eigen values, λ_i, and constants, A_1, are given as a function of heat transfer coefficients, h

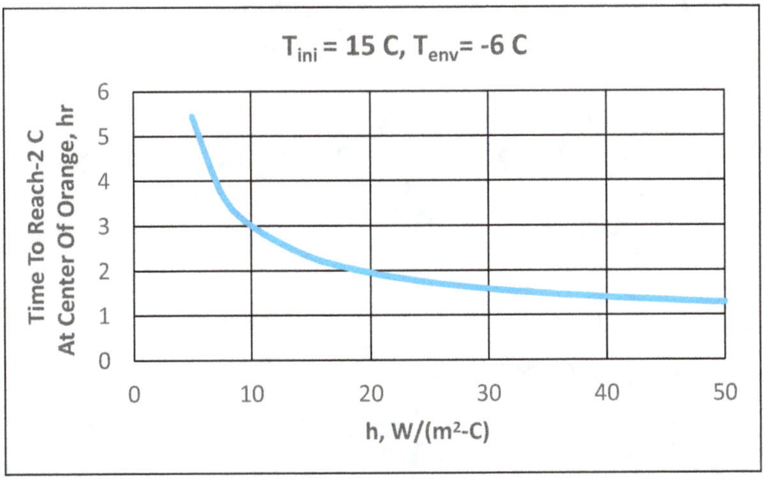

Figure 13-9 Time to reach -2 C at the center of orange as the heat transfer coefficient, h, at orange's surface increases

Absolute sensitivity of time to reach -2 C at the center of orange as the heat transfer coefficient, h, at orange's surface increase is decreasing, as shown in Figure 13-10. This absolute sensitivity, namely a change in time to reach -2 C at the center of orange to a change in the heat transfer coefficient, h, approach zero for $h > 50\ W/(m^2 - C)$.

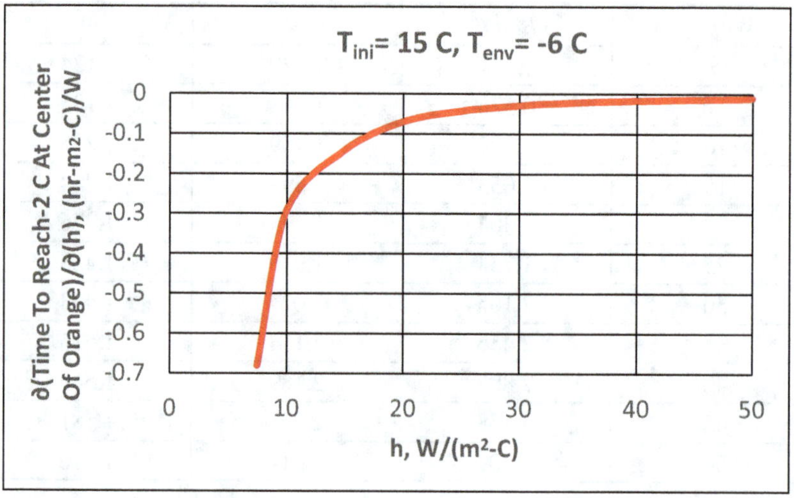

Figure 13-10 Sensitivity of time to reach -2 C at the center of orange as the heat transfer coefficient, *h*, at orange's surface increases

14

Stress Relieving a Stainless-Steel Plate in Shape of A Rectangular Parallelepiped

We will investigate cooling rate of a stainless-steel plate in shape of a rectangular parallelepiped during a stress relieving process as shown in Figure 14-1.

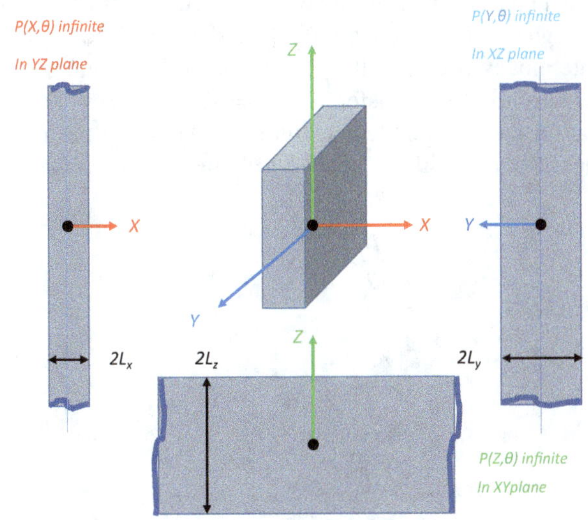

Figure 14-1 Stainless steel plate in rectangular parallelepiped shape. Its center, (0,0,0), is at intersecting centers of three infinite plates where finite dimensional views are shown.

For the three-dimensional transient temperature distribution of the rectangular parallelepiped, we will use product of three one-dimensional exact solutions for three infinite plates, namely one solution in the YZ plane for X-axis or P(X, θ), one solution in the XZ plane for the Y-axis or P(Y, θ), and one solution in the XY plane for the Z-axis or *P*(*Z*, θ).

An infinite plate in the YZ plane where the plate's thickness is much smaller than its width and height initially has a uniform temperature. Conduction heat transfer within the plate mostly occurs in the *X* direction. This infinite plate is suddenly subjected to convection heat transfer at its $X = \pm L_X$ surfaces. The partial differential equation for the transient temperature distribution in the X-axis with constant properties, *P*(*X*, θ), with its initial and boundary conditions are shown in Equations 14-1.

$$\frac{1}{\alpha}\frac{\partial T}{\partial \theta} = \frac{\partial^2 T}{\partial X^2} \qquad \text{14-1a}$$

$$T(X,0) = T_{ini} \qquad \text{14-1b}$$

$$\left(\frac{\partial T}{\partial X}\right)_{X=0} = 0 \qquad \text{14-1c}$$

$$-k\left(\frac{\partial T}{\partial X}\right)_{X=\pm L_X} = \bar{h}\left[T(\pm L_X,\theta) - T_{env}\right] \qquad \text{14-1d}$$

An infinite plate in the XZ plane where the plate's thickness is much smaller than its width and height has initially a uniform temperature. Conduction heat transfer within the plate mostly occurs in the *Y* direction. This infinite plate is suddenly subjected to convection heat transfer at its $Y = \pm L_Y$ surfaces. The partial differential equation for the transient temperature distribution in the Y-axis with constant properties, *P*(*Y*, θ), with its initial and boundary conditions are shown in Equations 14-2.

$$\frac{1}{\alpha}\frac{\partial T}{\partial \theta} = \frac{\partial^2 T}{\partial Y^2} \qquad \text{14-2a}$$

$$T(Y,0) = T_{ini} \qquad \text{14-2b}$$

$$\left(\frac{\partial T}{\partial Y}\right)_{Y=0} = 0 \qquad \text{14-2c}$$

$$-k\left(\frac{\partial T}{\partial Y}\right)_{Y=\pm L_Y} = \bar{h}\left[T(\pm L_Y,\theta) - T_{env}\right] \qquad \text{14-2d}$$

An infinite plate in the XY plane where the plate's thickness is much smaller than its width and height initially has a uniform temperature. Conduction heat transfer within the plate mostly occurs in the *Z* direction. This infinite plate is suddenly subjected to convection heat transfer at its $Z = \pm L_Z$ surfaces. The partial differential equation for the transient temperature distribution in the Z-axis with constant properties, $P(Z, \theta)$, with its initial and boundary conditions are shown in Equations 14-3.

$$\frac{1}{\alpha}\frac{\partial T}{\partial \theta} = \frac{\partial^2 T}{\partial Z^2} \quad \text{14-3a}$$

$$T(Z,0) = T_{ini} \quad \text{14-3b}$$

$$\left(\frac{\partial T}{\partial Z}\right)_{Z=0} = 0 \quad \text{14-3c}$$

$$-k\left(\frac{\partial T}{\partial Z}\right)_{Z=\pm L_Z} = \bar{h}\left[T(\pm L_Z,\theta) - T_{env}\right] \quad \text{14-3d}$$

In above equations, *T*, is the transient temperature in the specified directional location in meters and time θ in seconds. α is rectangular parallelepiped's thermal diffusivity in m^2/s. *k* is rectangular parallelepiped's thermal conductivity in W/(m-C). $\bar{h}$ is the forced air convection heat transfer coefficient averaged over surfaces of the rectangular parallelepiped in $W/(m^2\text{-}C)$. T_{ini} is the constant initial temperature of the whole rectangular parallelepiped in C. T_{env} is the constant temperature of the oncoming airstream to the rectangular parallelepiped during the cooling process in C. Walls of the stress relieving chamber have the same temperature as T_{env}.

Solutions to Equations 14-1 through to 14-3 are shown in Equations 14-4 through 14-6 for P(X,θ), P(Y,θ), P(Z,θ), respectively. See Ref. [15].

$$P(X,\theta) = \frac{T(X,\theta) - T_{env}}{T_{ini} - T_{env}} = \sum_{i=1}^{\infty} A_{i-X}\, exp^{\left(-\lambda_{i-X}^2 FO_X\right)} cos\left(\lambda_{i-X}\frac{X}{L_X}\right) \quad \text{14-4a}$$

$$Biot\,Number_X = \lambda_{i-X} tan\left(\lambda_{i-X}\right) \quad \text{14-4b}$$

$$A_{i-X} = \frac{4 sin\left(\lambda_{i-X}\right)}{2\lambda_{i-X} + sin\left(2\lambda_{i-X}\right)} \quad \text{14-4c}$$

$$P(Y,\theta) = \frac{T(Y,\theta) - T_{env}}{T_{ini} - T_{env}} = \sum_{i=1}^{\infty} A_{i-Y}\, exp^{\left(-\lambda_{i-Y}^2 FO_Y\right)} cos\left(\lambda_{i-Y}\frac{Y}{L_Y}\right) \quad \text{14-5a}$$

$$Biot\ Number_Y = \lambda_{i-Y} tan(\lambda_{i-Y}) \quad \text{14-5b}$$

$$A_{i-Y} = \frac{4sin(\lambda_{i-Y})}{2\lambda_{i-Y} + sin(2\lambda_{i-Y})} \quad \text{14-5c}$$

$$P(Z,\theta) = \frac{T(Z,\theta) - T_{env}}{T_{ini} - T_{env}} = \sum_{i=1}^{\infty} A_{i-Z} exp^{(-\lambda_{i-Z}^2 FO_Z)} cos\left(\lambda_{i-Z} \frac{Z}{L_Z}\right) \quad \text{14-6a}$$

$$Biot\ Number_Z = \lambda_{i-Z} tan(\lambda_{i-Z}) \quad \text{14-6b}$$

$$A_{i-Z} = \frac{4sin(\lambda_{i-Z})}{2\lambda_{i-Z} + sin(2\lambda_{i-Z})} \quad \text{14-6c}$$

The transient temperature distributions for the three-dimensional rectangular parallelepiped shown in Figure 14-1 is given in Equation 14-7.

$$\frac{T(X,Y,Z,\theta) - T_{env}}{T_{ini} - T_{env}} = P(X,\theta) \times P(Y,\theta) \times P(Z,\theta) \quad \text{14-7}$$

When $L_X \neq L_Y \neq L_Z$, eigen values λ_i and constants A_i have different values in $P(X,\theta)$, $P(Y,\theta)$, and $P(Z,\theta)$.

Biot number$_X$, Biot number$_Y$, and Biot number$_Z$ versus eigen values and constants, $\lambda_{i-X}, A_{i-X}, \lambda_{i-Y}, A_{i-Y}, \lambda_{i-Z}, A_{i-Z}$ are shown in Table 14-1 for $\bar{h} = 10\ W/(m^2 - C)$ and for i = 1, 2, 3, 4, *and* 5.

		λ_1	A_1	λ_2	A_2	λ_3	A_3	λ_4	A_4	λ_5	A_5
Biot #x	*0.278*	0.502	1.042	3.227	-0.052	6.327	0.014	9.454	-0.006	12.588	0.004
Biot #y	*0.417*	0.603	1.060	3.268	-0.074	6.349	0.020	9.469	-0.009	12.599	0.005
Biot #z	*0.556*	0.682	1.076	3.308	-0.095	6.370	0.027	9.489	-0.013	12.610	0.007

Table 14-1 Biot numbers versus eigen values, λ_i, and constants, A_i for i = *1,2,3,4, and 5* when $\bar{h} = 10\ W/(m^2 - C)$

We use following constant properties in Biot number and transient temperature distribution calculations.

$$k = 18\ W/(m - C); \qquad \alpha = 4 \times 10^{-6}\ m^2/s;$$

$$T_{ini} = 600\ C; \qquad T_{env} = 50\ C;$$

$$L_X = 0.5\ m; \qquad L_Y = 0.75\ m; \qquad L_Z = 1.0\ m$$

Series solutions given in Equations 14-4a, 14-5a and 14-6a can be approximated only by one term, i = 1, if the applicable Fourier number is greater than 0.2, $FO = \frac{\alpha\theta}{L^2} > 0.2$, as shown in Table 14-2.

$FO_X = \frac{\alpha\theta}{L_X^2} > 0.2$	$\theta > 12{,}500\,s$	$\theta > 3.5\,hr$
$FO_Y = \frac{\alpha\theta}{L_Y^2} > 0.2$	$\theta > 28{,}125\,s$	$\theta > 7.8\,hr$
$FO_Z = \frac{\alpha\theta}{L_Z^2} > 0.2$	$\theta > 50{,}000\,s$	$\theta > 13.9\,hr$

Table 14-2 Series solutions given in Equations 14-4a, 14-5a and 14-6a can be approximated only by one term, i = 1, if times are greater than shown values.

Stress relieving of the present rectangular parallelepiped stainless steel piece will be done by forced air cooling in order to expedite the cooling process from 600 C down to 100 C. Also, during the stress relieving process, engineers would like to minimize distortion so dimensions can stay within their tolerances. Maximum allowable average temperature change with respect to time during stress relieving is given in Equation 14-8.

$$\left(\frac{600-100)}{\Delta\theta}\right)_{average} = \left(0.5\ C/min\right)_{maximum} \qquad 14\text{-}8$$

Since, we are interested in finding the maximum allowable average temperature change with respect to time, Equation 14-8, during forced air cooling, we cannot use one term approximations to Equations 14-4a, 14-5a and 14-6a. We have to investigate temperature change with respect to time in the stainless-steel plate from $\theta = 0\,s$ to times when locations in the plate reach 100 C. Therefore, we will use five terms, i = 1 *through* 5, in our temperature versus time calculations for more accurate results from Equations 14-4a, 14-5a and 14-6a.

Temperature changes with respect to time at three locations of the rectangular parallelepiped are shown in Figure 14-2 for forced air convection heat transfer coefficient of $\overline{h} = 10W/\left(m^2 - C\right)$. Maximum average temperature change with respect to time while cooling from 600 C to 100 C occurs at eight corners of the

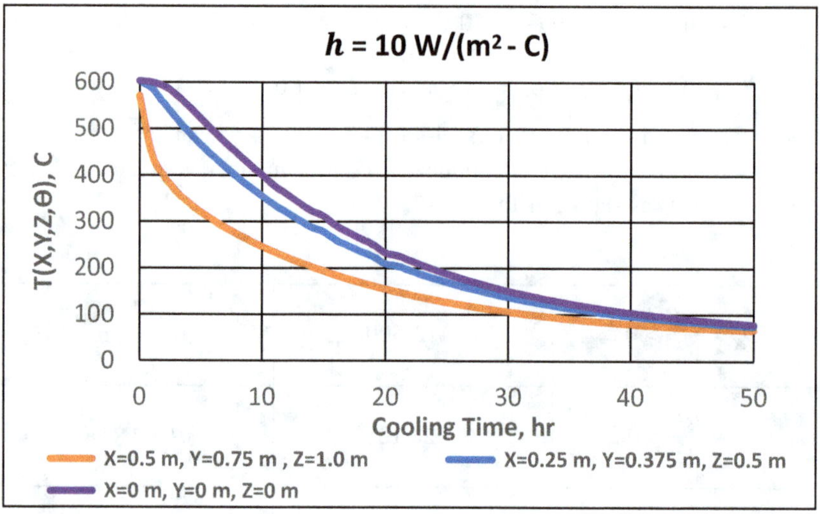

Figure 14-2 Temperature-time profile at three locations of the rectangular parallelepiped

stainless-steel piece, namely at $X = \pm 0.5\,m,\ Y = \pm 0.75\,m,\ Z = \pm 1.0\,m$. The minimum average temperature change with respect to time occurs at its center at $X = 0\,m, Y = 0\,m,\ Z = 0\,m.$

Maximum average temperature change with respect to time while cooling from 600 C to 100 C occurs at eight corners of the stainless-steel piece which can also be obtained from the derivative of temperature, $T(X,Y,Z,\theta)$, with respect to time, θ, in Equation 14-7 as shown in Equation 14-9.

$$\frac{\partial T\left(X,Y,Z,\theta\right)}{\partial \theta} = \left(T_{ini} - T_{env}\right)\times$$

$$\left(\frac{-\alpha}{L_X^2}\right)\left[\sum_{i=1}^{\infty} A_{i-X}\lambda_{i-X}^2 exp^{-\lambda_{i-X}^2 FO_X} cos\left(\lambda_{i-X}\frac{X}{L_X}\right)\right]\times P\left(Y,\theta\right)\times P\left(Z,\theta\right)\times$$

$$\left(\frac{-\alpha}{L_Y^2}\right)\left[\sum_{i=1}^{\infty} A_{i-Y}\lambda_{i-Y}^2 exp^{-\lambda_{i-Y}^2 FO_Y} cos\left(\lambda_{i-Y}\frac{Y}{L_Y}\right)\right]\times P\left(X,\theta\right)\times P\left(Z,\theta\right)\times \quad 14\text{-}9$$

$$\left(\frac{-\alpha}{L_Z^2}\right)\left[\sum_{i=1}^{\infty} A_{i-Z}\lambda_{i-Z}^2 exp^{-\lambda_{i-Z}^2 FO_Z} cos\left(\lambda_{i-Z}\frac{Z}{L_Z}\right)\right] P\left(X,\theta\right)\times P\left(Y,\theta\right)$$

Maximum average temperature changes with respect to time while cooling from 600 C to 100 C at eight corners of the stainless-steel piece versus forced air convection heat transfer coefficient is shown in Figure 14-3. During stress relieving,

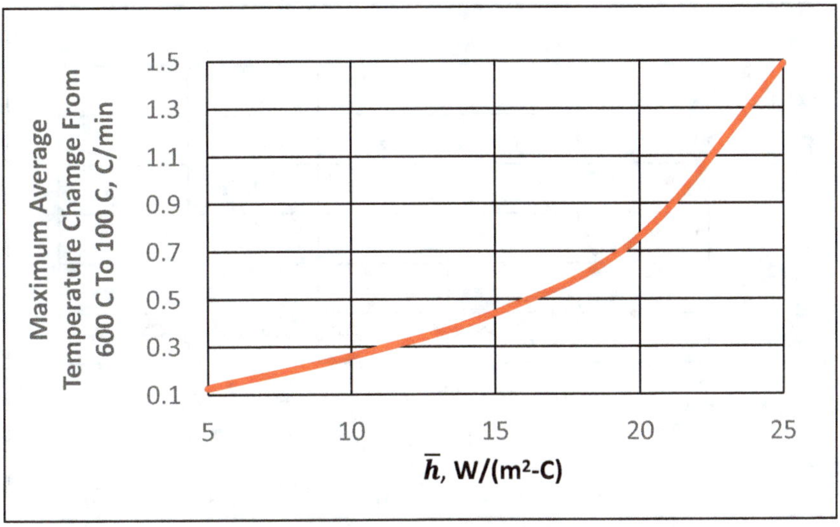

Figure 14-3 Maximum average temperature changes with respect to time while cooling from 600 C to 100 C at eight corners of the stainless-steel piece versus forced air convection heat transfer coefficient

required slow cooling of 0.5 C/min, occurs at $\bar{h} = 16\ W/\left(m^2 - C\right)$ as can be observed in Figure 14-3.

Temperature profiles versus time while cooling from 600 C at eight corners of the stainless-steel rectangular parallelepiped for various average forced air convection heat transfer coefficients for air from 5 to 25 W/(m^2 – C) are shown in Figure 14-4. When $\bar{h} = 5\ W/\left(m^2 - C\right)$, it takes 66.85 hours for eight corners of the stainless-steel rectangular parallelepiped to cool from 600 C to 100 C. In this case, the average cooling rate for these corners is 0.126 C/min. When $\bar{h} = 25\ W/\left(m^2 - C\right)$, it takes 5.61 hours for eight corners of the stainless-steel rectangular parallelepiped to cool from 600 C to 100 C. In this case, the average cooling rate for these eight corners is 1.49 C/min.

Average forced air convection heat transfer coefficient for cross flow over a non-circular cylinder shape, with an air velocity upstream of the rectangular parallelepiped, Reynolds number, $Re_D \dfrac{VD}{\nu}$, is determined from the empirical correlation by Hilpert, R., given in Equation 14-10, See Ref. [9]. Equivalent diameter, D, to a circular cylinder is obtained from, $\dfrac{\pi D^2}{4} = 2L_X \times 2L_Z$.

$$\overline{Nu} = \frac{\bar{h}D}{k_{air}} = 0.027 Re_D^{0.805} Pr^{0.333} \qquad 14\text{-}10$$

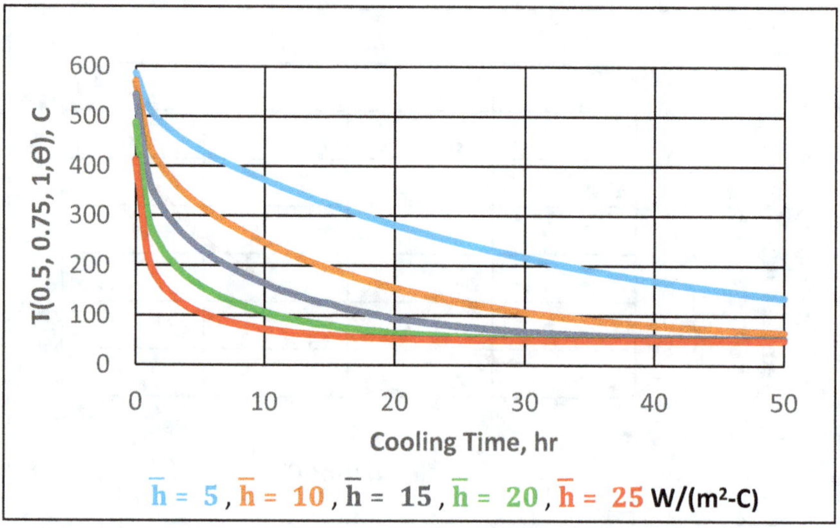

Figure 14-4 Temperature profiles versus time while cooling from 600 C at eight corners of the stainless-steel rectangular parallelepiped for various average forced air convection heat transfer coefficients

Air properties such as, thermal conductivity, k_{air}, kinematic viscosity, v_{air}, Prandtl number, Pr_{air}, in Equation 14-10 are taken at the film temperature of $0.5(50+600)=325\ C$ and at sea level pressure. During stress relieving of the stainless-steel plate in rectangular parallelepiped shape, the required maximum averaged cooling rate from 600 C to 100 C at 0.5 C/min, occurs when the average heat transfer coefficient is $\bar{h}=16\ W/\left(m^2-C\right)$, and requires an upstream air velocity at 5.60 m/s as shown in Figure 14-5.

When the average heat transfer coefficient is $\bar{h}=16\ W/\left(m^2-C\right)$, with an upstream air velocity at 5.60 m/s, eight corners of the stainless-steel plate in rectangular parallelepiped shape with $L_X=0.5m, L_Y=0.75m, L_Z=1.0m$ will cool down from 600 C down to 100 C in 17.05 hours within the maximum allowable average cooling rate of 0.5 C/min.

The minimum average temperature change with respect to time occurs at its center at $X=0m,\ Y=0m,\ Z=0m$. The center of the stainless-steel plate, T(0, 0, 0, θ) cools down from 600 C to 100 C in 26.8 hours as shown in Figure 14-6. The average cooling rate at the center is 0.31 C/min which is 38% less than the maximum allowable average cooling rate of 0.5 C/min.

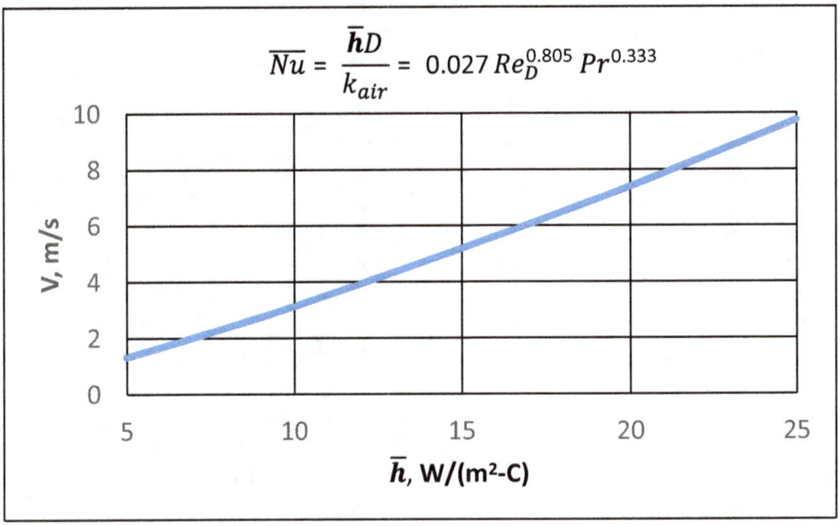

Figure 14-5 Upstream air velocity versus average convection heat transfer coefficient obtained from Equation 14-10

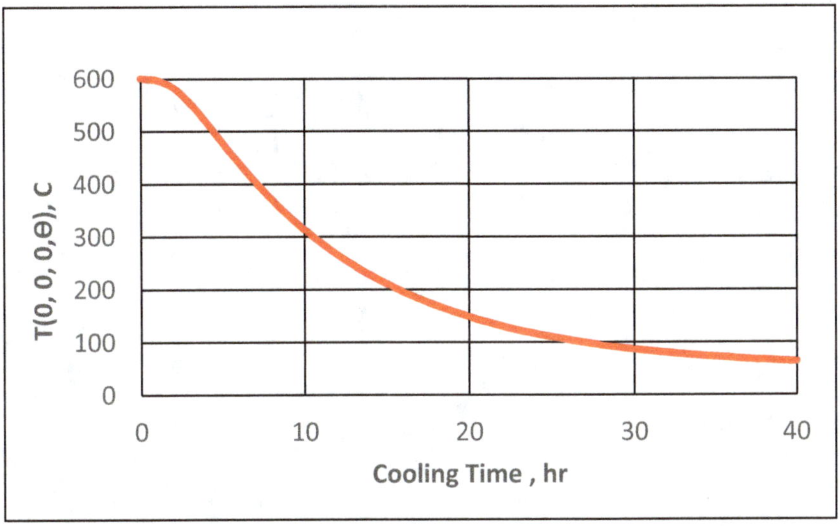

Figure 14-6 Temperature profiles versus time while cooling from 600 C at the center of the stainless-steel rectangular parallelepiped for average forced air convection heat transfer coefficient of 16 W/(m^2 – C)

15

Heat Transfer Between Two Semi-Infinite Solids in Contact

We will investigate heat transfer between two semi-infinite solids when they are brought into thermal contact. One-dimensional transient conduction heat transfer in a semi-infinite solid has been solved with different boundary conditions in most heat transfer books. For example: See Ref. [10]. There is a constant surface temperature applied to $X = 0$ surface, see Figure 15-1, which is different than semi-infinite solid's initial temperature. Following partial differential equation and initial and boundary conditions apply as given in Equations 15-1.

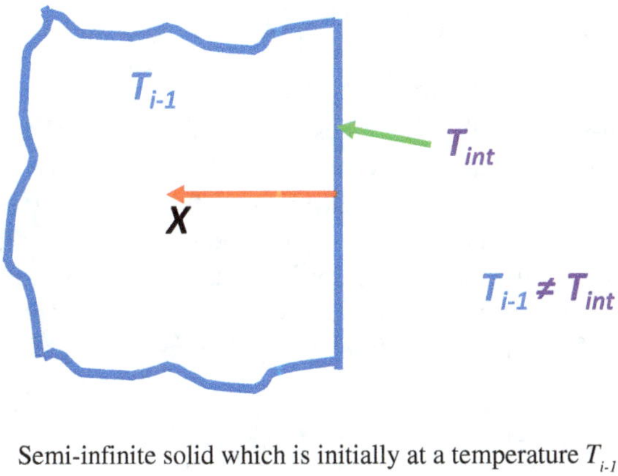

Figure 15-1 Semi-infinite solid which is initially at a temperature T_{i-1} and which is exposed to a constant surface temperature T_{int} at all times and $\boldsymbol{T_{i-1} \neq T_{int}}$

$$\frac{\partial^2 T}{\partial X^2} = \frac{1}{\alpha}\frac{\partial T}{\partial \theta} \qquad \text{15-1a}$$

$$T(X,0) = T_{i-1} \qquad \text{15-1b}$$

$$T(X \to \infty, \theta) = T_{i-1} \qquad \text{15-1c}$$

$$T(0,\theta) = T_{int} \qquad \text{15-1d}$$

$T(X,\theta)$ is the temperature of the semi-infinite solid at the X location in meters from its surface and at a time θ in seconds. The temperature distribution in a semi-infinite solid that satisfies Equations 15-1 is shown in Equation 15-2. The heat transfer at the surface of the semi-infinite solid, $X = 0$, is given in Equation 15-3.

$$\frac{T(X,\theta) - T_{int}}{T_{i-1} - T_{int}} = erf\left(\frac{X}{2\sqrt{\alpha\theta}}\right) \qquad \text{15-2}$$

$$q(0,\theta) = \frac{k\left(T_{i-1} - T_{int}\right)}{\sqrt{\pi\alpha\theta}} \qquad \text{15-3}$$

where k is semi-infinite solid's thermal conductivity in W/(m-K) and α is semi-infinite solid's thermal diffusivity in m²/s. $\left(\frac{X}{2\sqrt{\alpha\theta}}\right)$ is the dimensionless similarity variable that converts the partial differential Equation 15-1a into an ordinary one. $erf\left(\frac{X}{2\sqrt{\alpha\theta}}\right)$ is the Gaussian error function defined in Equation 15-4 and plotted below in Figure 15-2.

$$erf\left(\frac{X}{2\sqrt{\alpha\theta}}\right) = \frac{2}{\sqrt{\pi}}\int_0^{\frac{X}{2\sqrt{\alpha\theta}}} exp\left(-\xi^2\right)d\xi \qquad \text{15-4}$$

When two semi-infinite solids' surfaces at two different initial temperatures are brought into contact, heat flux leaving one solid is equal to the heat flux going into the second solid. Also, the interface temperature between two semi-infinite solids, T_{int}, that are in contact without any contact resistance, should be the same. Using Equation 15-3, the interface temperature can be determined from balanced heat fluxes as shown in Equations 15-5 where X-direction for $q_{1\to int}(0,\theta)$ is pointing to the left and for $q_{int\to 2}(0,\theta)$ it is pointing to the right.

$$q_{1\to int}(0,\theta) = -\frac{k_1\left(T_{int} - T_{i-1}\right)}{\sqrt{\pi\alpha_1\theta}} = \frac{k_2\left(T_{int} - T_{i-2}\right)}{\sqrt{\pi\alpha_2\theta}} = q_{int\to 2}(0,\theta) \qquad \text{15-5a}$$

$$T_{i-2} < T_{int} < T_{i-1} \qquad \text{15-5b}$$

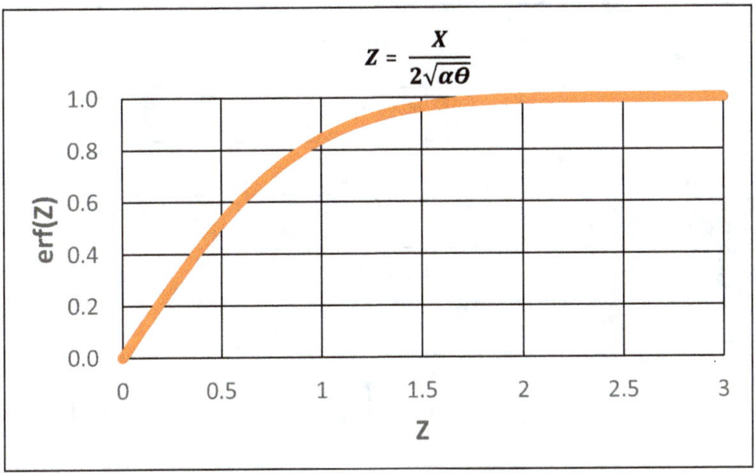

Figure 15-2 Gaussian error function as a function of $Z = \dfrac{X}{2\sqrt{\alpha\theta}}$

The interface temperature between two solids, T_{int}, can be obtained from Equation 15-5a and it is shown in Equation 15-6.

$$T_{int} = \frac{\left(k_1 / \sqrt{\alpha_1}\right) T_{i-1} + \left(k_2 / \sqrt{\alpha_2}\right) T_{i-2}}{\left(k_1 / \sqrt{\alpha_1}\right) + \left(k_2 / \sqrt{\alpha_2}\right)} \qquad 15\text{-}6$$

The interface contact temperature between two solids, T_{int}, depends on thermal properties k and α of two solids and their initial temperature, T_{i-1} and T_{i-2} at all times. The interface contact temperature between two solids, T_{int}, is shown in Figure 15-3 for $T_{i-1} = 100\ C$ and $T_{i-2} = 0\ C$ as a function of $\dfrac{k_1 / \sqrt{\alpha_1}}{k_2 / \sqrt{\alpha_2}}$.

When $\dfrac{k_1 / \sqrt{\alpha_1}}{k_2 / \sqrt{\alpha_2}} = 1$, $T_{int} = \dfrac{T_{i-1} + T_{i-2}}{2} = 50C$, the average of two semi-infinite solid's initial temperatures. If $k_1 / \sqrt{\alpha_1} > k_2 / \sqrt{\alpha_2}$, T_{int} approaches T_{i-1}. If $k_2 / \sqrt{\alpha_2} > k_1 / \sqrt{\alpha_1}$, T_{int} approaches T_{i-2}.

Heat transfer between two contacting semi-infinite solids is given in Equation 15-7 as functions of their thermal properties and their initial temperatures.

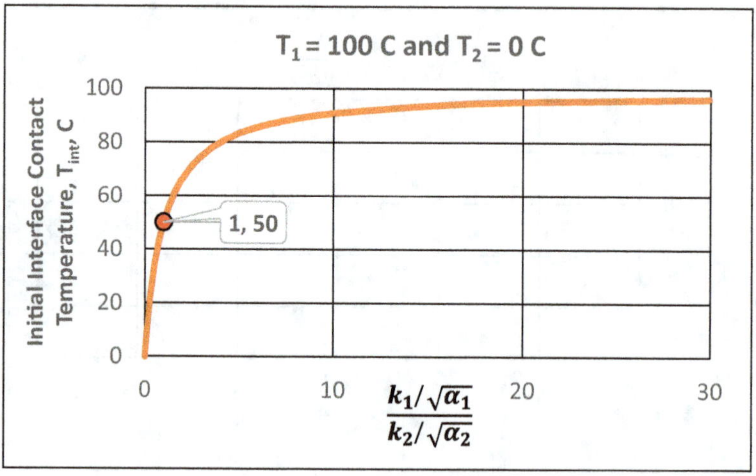

Figure 15-3 T_{int} versus $\dfrac{k_1/\sqrt{\alpha_1}}{k_2/\sqrt{\alpha_2}}$ for $T_{i-1} = 100C$ and $T_{i-2} = 0C$

$$q_{1\rightarrow int}(0,\theta) = q_{int\rightarrow 2}(0,\theta) = \frac{\left(k_1/\sqrt{\alpha_1}\right)\times\left(k_2/\sqrt{\alpha_2}\right)}{\left(k_1/\sqrt{\alpha_1}\right)+\left(k_2/\sqrt{\alpha_2}\right)}\times\frac{\left(T_{i-1}-T_{i-2}\right)}{\sqrt{\pi\theta}} \qquad \text{15-7}$$

Now, let us investigate how heat transfers from semi-infinite solid, 1 which is making thermal contact to semi-infinite solid, 2. Let us rewrite Equation 15-7 by introducing a ***heat transfer enhancement factor*** by using thermal properties of semi-infinite solid 1 and semi-infinite solid 2, as shown in Equation 15-8.

$$q_{1\rightarrow int}(0,\theta) = \left[\frac{1}{\dfrac{\left(k_1/\sqrt{\alpha_1}\right)}{\left(k_2/\sqrt{\alpha_2}\right)}+1}\right]\times\left[\frac{k_1\left(T_{i-1}-T_{i-2}\right)}{\sqrt{\pi\alpha_1\theta}}\right] \qquad \text{15-8}$$

The first bracket includes the heat transfer enhancement factor by semi-infinite solid 1's and semi-infinite solid 2's thermal properties. If $\left(k_2/\sqrt{\alpha_2}\right)$ is much larger than $\left(k_1/\sqrt{\alpha_1}\right)$, the heat transfer enhancement factor approaches unity and semi-infinite solid 2 does not inhibit heat transferring from semi-infinite solid 1 while making a thermal contact, namely $T_{int} \rightarrow T_{i-2}$. If $\left(k_2/\sqrt{\alpha_2}\right)$ is much smaller than $\left(k_1/\sqrt{\alpha_1}\right)$, then, the heat transfer enhancement factor approaches zero or $T_{int} \rightarrow T_{i-1}$ and semi-infinite solid 2 does inhibit heat transferring from semi-infinite solid 1 while making a thermal contact. The second bracket represents the heat transfer at the surface of semi-infinite solid 1 when its surface temperature is a constant at T_{i-2}. Equation 15-8 also verifies that when a human touches an object, the feeling of hot or cold depends on the heat transfer between our body and the other object. Higher heat transfer between two bodies increases our sense of hot or cold between our body and the other object.

Let us touch three different objects, assumed to be semi-infinite solids, at the same temperature of 25 C. The question is which one will feel the coldest to our touch when our hand is at 37 C. Thermal properties of our hand and three objects are given in Table 15-1.

Higher the values of, $k/\sqrt{\alpha} = \sqrt{k\rho C_p}$, of a solid that we touch means that conduction heat transfer rate times the volumetric heat capacity of that solid enhances heat transfer from our hand to the solid and the surface of the solid feels colder to us. Heat transfer enhancement factors, $\dfrac{1}{\dfrac{\left(k_1/\sqrt{\alpha_1}\right)}{\left(k_2/\sqrt{\alpha_2}\right)}+1}$, as defined in Equation 15-8, for touching pine wood, pure gold and granite stone with our hand are shown in Table 15-2.

The coldest feeling comes from touching pure gold's surface with our hand, since 96% of available heat flux from our hand to a constant temperature surface at, T_{i-2} flows out. A lesser cold feeling comes from touching granite stone's surface with our hand, since only 66% of available heat flux from our hand to a constant temperature surface at, T_{i-2}, flows out.

The minimum cold feeling comes from touching pine wood's surface with our hand, since only 27% of available heat flux from our hand to a constant temperature surface at, T_{i-2} flows out.

Heat transfer from semi-infinite solid, 1, at an initial temperature of T_{i-1}, which is making thermal contact to semi-infinite solid, 2, at an initial temperature of

	T_i, C	k, W/(m-C)	α, m²/s	$k/\sqrt{\alpha}$, J/(m²-C-s$^{0.5}$)
Human Hand	37	0.37	8.92×10^{-8}	1,239.0
Pine Wood	25	0.12	6.91×10^{-8}	456.4
Pure Gold	25	317	1.27×10^{-4}	28,093.3
Granite Stone	25	2.79	1.37×10^{-6}	2,384.7

Table 15-1 Thermal properties of our hand, pine wood, pure gold, and granite stone

	Human Hand Touches Pine Wood	Human Hand Touches Pure Gold	Human Hand Touches Granite Stone
Heat Transfer Enhancement Factor	0.27	0.96	0.66

Table 15-2 The heat transfer enhancement factors for touching pine wood, pure gold and granite stone with our hand

T_{i-2}, versus $\dfrac{\left(k_1/\sqrt{\alpha_1}\right)}{\left(k_2/\sqrt{\alpha_2}\right)}$ is shown in Figure 15-4. When two solids have the same thermal properties, like shaking hands with someone, and if there is a 1 C difference between two hands, $T_{i-1} > T_{i-2}$, 50% of the heat transfer is $q_{1\to int}(0,\theta)$ and the other 50% is $q_{int\to 2}(0,\theta)$.

For values where, $k_1/\sqrt{\alpha_1} > k_2/\sqrt{\alpha_2}$, means that the conduction heat transfer rate times the volumetric heat capacity of that solid 2 is less than the conduction

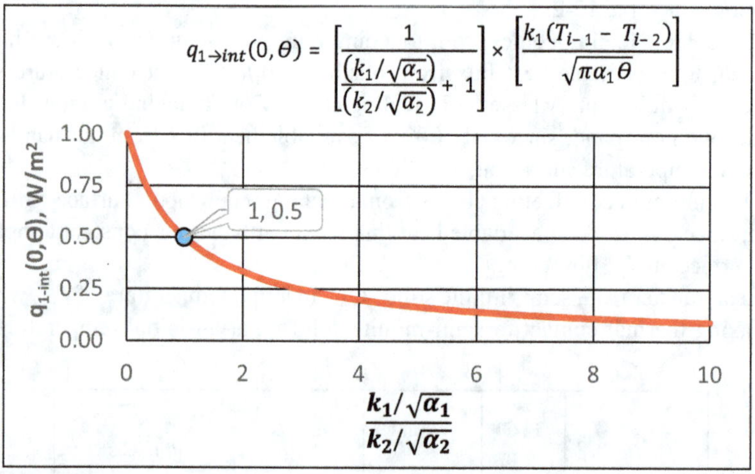

Figure 15-4 Heat transfer from semi-infinite solid, 1, at an initial temperature of T_{i-1}, which is making thermal contact to semi-infinite solid, 2, at an initial temperature of T_{i-2}, versus $\dfrac{\left(k_1/\sqrt{\alpha_1}\right)}{\left(k_2/\sqrt{\alpha_2}\right)}$

heat transfer rate times the volumetric heat capacity of solid 1. Therefore, T_{int}, approaches T_{i-1} while reducing $q_{1\rightarrow int}(0,\theta)$.

For values where, $k_1/\sqrt{\alpha_1} < k_2/\sqrt{\alpha_2}$, means that the conduction heat transfer rate times the volumetric heat capacity of that solid 2 is more than the conduction heat transfer rate times the volumetric heat capacity of solid 1. Therefore, T_{int}, approaches T_{i-2} while increasing $q_{1\rightarrow int}(0,\theta)$.

Sensitivity of T_{int} to changes in $\dfrac{(k_1/\sqrt{\alpha_1})}{(k_2/\sqrt{\alpha_2})}$ is shown in Figure 15-5.

As $k_1/\sqrt{\alpha_1} > k_2/\sqrt{\alpha_2}$, T_{int} approaches T_{i-1} and its sensitivity to changes in $\dfrac{(k_1/\sqrt{\alpha_1})}{(k_2/\sqrt{\alpha_2})}$ decreases. As $k_1/\sqrt{\alpha_1} > k_2/\sqrt{\alpha_2}$, T_{int} approaches T_{i-2} and its sensitivity to changes in $\dfrac{(k_1/\sqrt{\alpha_1})}{(k_2/\sqrt{\alpha_2})}$ increases.

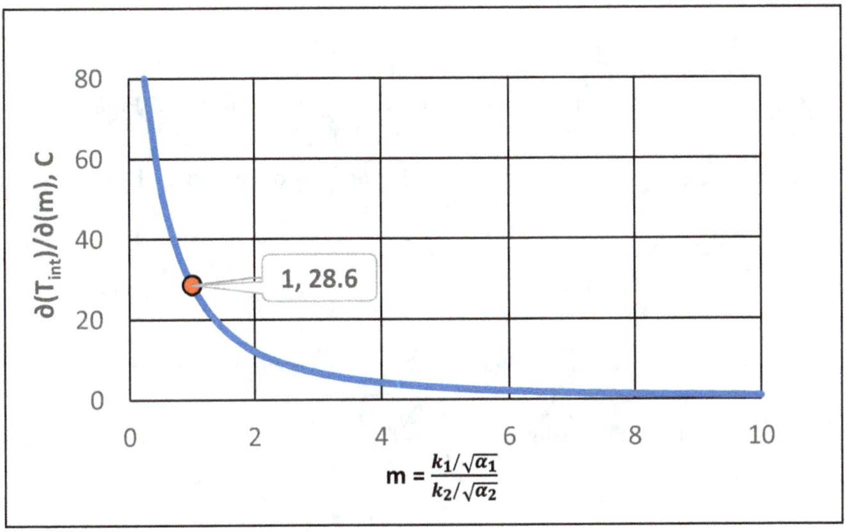

Figure 15-5 $\partial(T_{int})/\partial(m)$ versus $m = \dfrac{(k_1/\sqrt{\alpha_1})}{(k_2/\sqrt{\alpha_2})}$

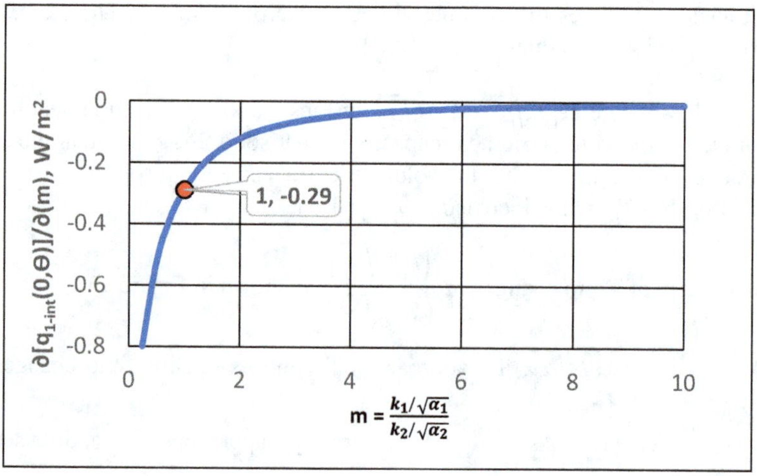

Figure 15-6 $\partial[q_{1\text{-int}}(0,\theta)]/\partial(m)$ versus $m = \dfrac{\left(k_1/\sqrt{\alpha_1}\right)}{\left(k_2/\sqrt{\alpha_2}\right)}$

Sensitivity of $q_{1\rightarrow int}(0,\theta)$ to changes in $\dfrac{\left(k_1/\sqrt{\alpha_1}\right)}{\left(k_2/\sqrt{\alpha_2}\right)}$ is shown in Figure 15-6. As $k_1/\sqrt{\alpha_1} < k_2/\sqrt{\alpha_2}$, T_{int} approaches $T_{i\text{-}1}$ and absolute value of changes to decreases to changes in $\dfrac{\left(k_1/\sqrt{\alpha_1}\right)}{\left(k_2/\sqrt{\alpha_2}\right)}$.

As $k_1/\sqrt{\alpha_1} < k_2/\sqrt{\alpha_2}$, T_{int} approaches $T_{i\text{-}2}$ and absolute value of changes to $q_{1\rightarrow int}(0,\theta)$ increases to changes in $\dfrac{\left(k_1/\sqrt{\alpha_1}\right)}{\left(k_2/\sqrt{\alpha_2}\right)}$.

16

Heat Transfer Between a Log Cabin and a Brush Fire

We will investigate heat transfer between a log cabin and a surrounding brush fire. We will assume the pine wood logs of the cabin to be semi-infinite solids. One-dimensional transient conduction heat transfer in a semi-infinite solid with combined convection and radiation heat transfers at pine wood log cabins surface will be used. There is a constant surface convection and radiation heat transfer from burning brush applied suddenly at $X = 0$ surface as shown in Figure 16-1. Applicable partial differential equation and initial and boundary conditions are given in Equations 16-1.

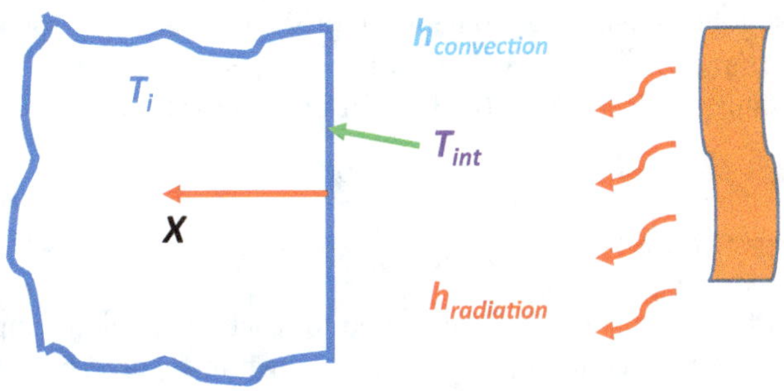

Figure 16-1 Semi-infinite solid which is initially at a temperature T_i and which is exposed suddenly to a constant surface convection and radiation heat transfer from burning brush applied at $X = 0$ surface

$$\frac{\partial^2 T}{\partial X^2} = \frac{1}{\alpha}\frac{\partial T}{\partial \theta} \quad \text{16-1a}$$

$$T(X,0) = T_i \quad \text{16-1b}$$

$$(X \to \infty, \theta) = T_i \quad \text{16-1c}$$

$$\left(-k\frac{\partial T}{\partial X}\right)_{X=0} = h_{total}\left[T_{sur} - T(0,\theta)\right] \quad \text{16-1d}$$

$T(X,\theta)$ is the temperature of the semi-infinite solid in K or in C at the X location in meters from its surface and at a time θ in seconds. The transient temperature distribution in a semi-infinite solid that satisfies Equations 16-1 is shown in Equation 16-2. See Ref. [15] for details.

$$\frac{T(X,\theta)-T_i}{T_{sur}-T_i} = erfc\left(\frac{X}{2\sqrt{\alpha\theta}}\right) - \left[exp\left(\frac{h_{total}X}{k} + \frac{h_{total}^2\alpha\theta}{k^2}\right)\right]\left[erfc\left(\frac{X}{2\sqrt{\alpha\theta}} + \frac{h_{total}\sqrt{\alpha\theta}}{k}\right)\right] \quad \text{16-2}$$

where k is semi-infinite solid's thermal conductivity in W/(m-K) and α is semi-infinite solid's thermal diffusivity in m^2/s. $erfc(m) = 1 - erf(m)$ is the complementary Gaussian error function. $erf(m)$ defined in Chapter 15, Equation 15-4 and plotted in Figure 15-2. h_{total} is the time-averaged combined convection and radiation heat transfer coefficient applied as a constant to the pine wood's surface. h_{total} varies with time due to changing fire conditions, increasing surface temperature and changing wind conditions. Therefore, $5 < h_{total} < 50W/(m^2-K)$ will be used as a time-averaged constant to determine times for pine wood's surface to reach its ignition temperature of 350 C (623 K).

Let us use following constant pine wood properties and its surrounding properties to determine the time it takes for surface temperature of pine wood to reach its ignition temperature of 350 C (623 K) for different values of h_{total}.

$T_i = 25C;$ $\quad$ $T_{sur} = 600C$

$k = 0.12W/(m-K);$ $\quad$ $\alpha = 8.35\times10^{-8}\,m^2/s$

Times for surface temperature of pine wood to reach its ignition temperature of 350 C (623 K) for different h_{total} values are shown in Figure 16-2. h_{total} varies between 5 W/(m^2-K) to 50 W/(m^2-K). When the heat transfer coefficient is only due to air convection, about 5 W/(m^2-K), it takes the pine wood's surface 109.1 minutes to reach 350 C. When radiation heat transfer dominates the total heat transfer coefficient at about 50 W/(m^2-K), it takes the pine wood surface 1.1 minutes to reach 350 C.

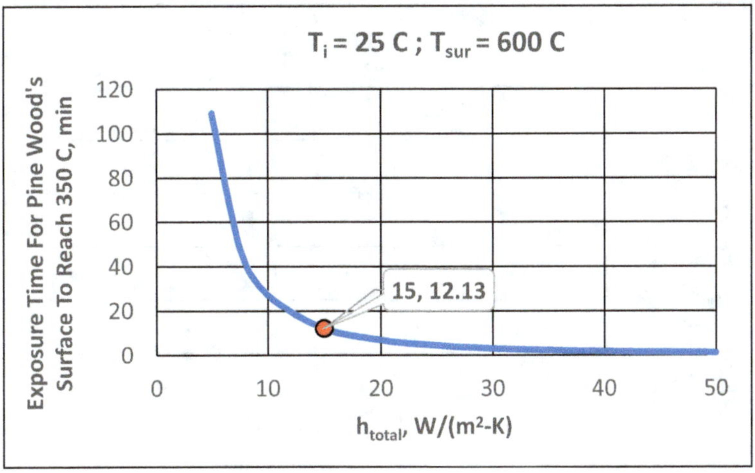

Figure 16-2 Times for surface temperature of pine wood to reach its ignition temperature of 350 C (623 K) at $X = 0$ for different h_{total} values. $\frac{h_{total}}{k}\sqrt{\alpha\theta} = 0.974$

Times for surface temperature of pine wood to reach its ignition temperature of 350 C (623 K) at $X = 0$ for different h_{total} values vary as shown in Equation 16-3. Equation 16-3 is calculated from Equation 16-2 for $X = 0$.

$$\frac{h_{total}}{k}\sqrt{\alpha\theta} = 0.974 \qquad 16\text{-}3$$

Exposure times for pine wood's surface to reach 350 C versus the total heat transfer coefficient decreases fast at low total heat transfer coefficient values while every other parameter is a constant as given above.

Exposure times approach a minute or less when $h_{total} > 35\ W/\left(m^2 - K\right)$. This analysis is done while other parameters such as T_i, T_{sur}, k, and α are constants as specified above.

Change in exposure time for pine wood's surface, $X = 0$, to reach 350 C divided by change in total heat transfer coefficient versus total heat transfer coefficient is shown in Figure 16-3. This is the slope of the curve that was presented above in Figure 16-2 and the equation for it is shown in Equation 16-4.

$$\frac{\partial\theta}{\partial h_{total}} = -1.9\frac{k^2}{\alpha h_{total}^3} \qquad 16\text{-}4$$

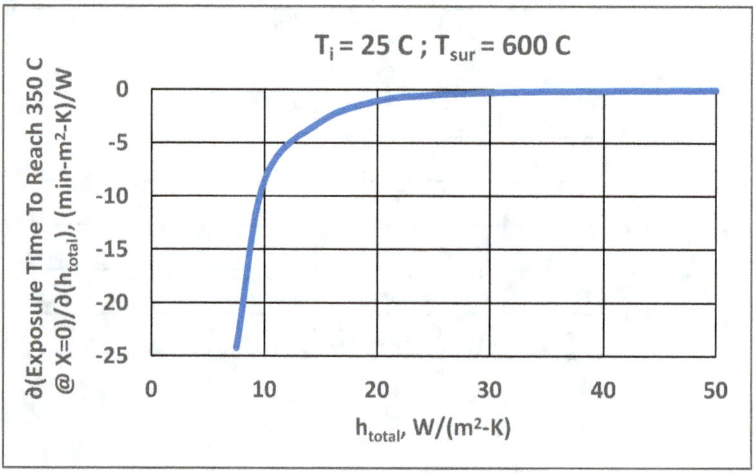

Figure 16-3 Change in exposure time for pine wood's surface to reach 350 C divided by change in total heat transfer coefficient versus total heat transfer coefficient

Absolute values of change in exposure time to change in total heat transfer coefficient are high when $h_{total} > 20\ W/\left(m^2 - K\right)$. Sensitivity of change in exposure time to change in total heat transfer coefficient approaches zero when $h_{total} > 20\ W/\left(m^2 - K\right)$ and it varies as $\frac{-1}{h_{total}^3}$.

With time averaged viewing factors of about 0.3 between the brush fire and log cabin's pine wood surface, a total heat transfer coefficient of 15 W/(m^2-K) is assumed for following calculations. At this total heat transfer coefficient value, it takes the pine wood surface 12.13 minutes to reach 350 C.

Temperature-time distributions for $X = 0\,m, 0.01\,m, 0.02\,m,\ and\ 0.05\,m$ for a total heat transfer coefficient of 15 W/(m^2-K) are shown in Figure 16-4. As expected, the surface temperature distribution is most affected by surrounding conditions as shown in Equation 16-2. As we go deeper into the semi-infinite solid, temperatures stay at the initial temperature of $T_i = 25\ C$ longer and longer, i.e 21 minutes at. $X = 0.05\ m$ in Figure 16-4. As time increases, temperature response to surface conditions at deeper locations depend on material properties such as the thermal conductivity, k, and the thermal diffusivity, α. For example, these two material properties for pine wood are three orders of magnitude less than those for aluminum.

Spatial temperature distributions for $\theta = 15\,min, 30\ min, 45\,min,\ and\ 60\,min$ and for a total heat transfer coefficient of 15 W/(m^2-K) are shown in Figure 16-5. Heat penetrates into the pine wood as time progresses. It takes 15 minutes for thermal energy to penetrate into a depth of 4 cm.

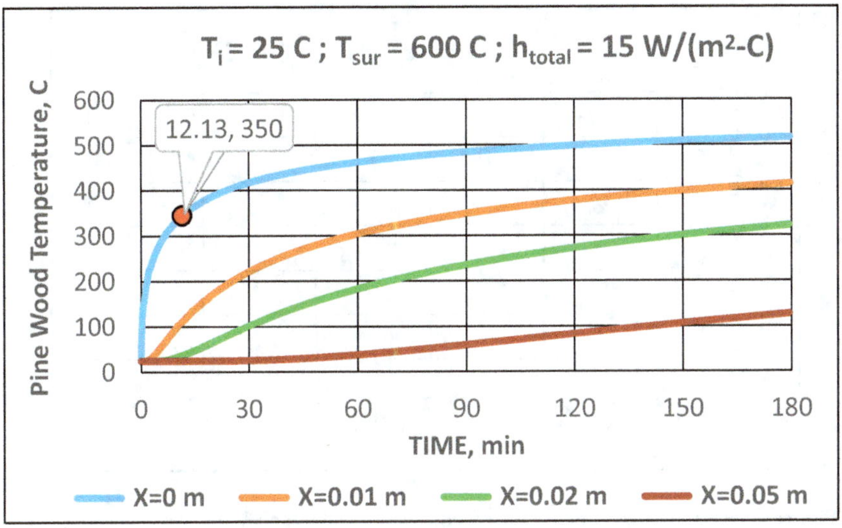

Figure 16-4 Transient temperature distributions for $X = 0\,m, 0.01\,m, 0.02\,m,\ and\ 0.05\,m$ for a total heat transfer coefficient of 15 W/(m²-K)

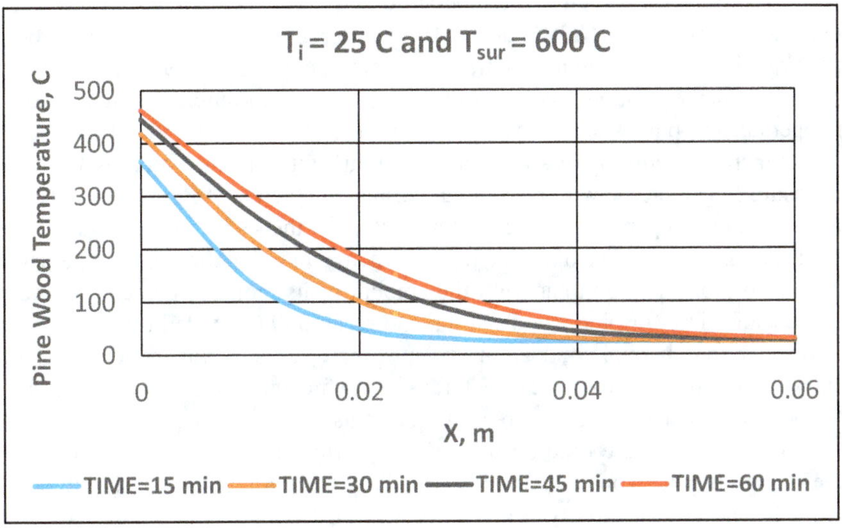

Figure 16-5 Spatial temperature distributions in pine wood at different times for a total heat transfer coefficient of 15 W/(m²-K)

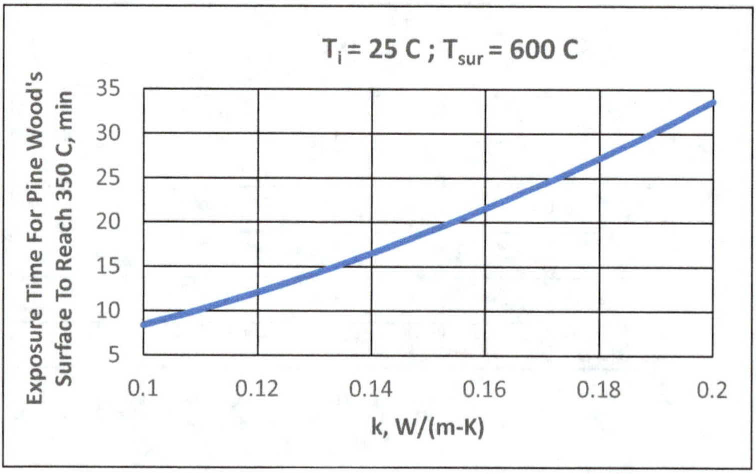

Figure 16-6 Exposure time for pine wood's surface to reach 350 C versus pine wood's thermal conductivity for a total heat transfer coefficient of 15 $W/\left(m^2-K\right)$

Slopes of spatial temperature distributions indicate the amount of conduction heat transfer at that time and location.

Exposure times for surface temperature of pine wood to reach its ignition temperature of 350 C (623 K) at $X = 0$ for different k is shown in Figure 16-6 per Equation 16-3 for a total heat transfer coefficient of 15 $W/\left(m^2-K\right)$ and for other parameters that are given above as constants. Exposure times for surface temperature of pine wood to reach its ignition temperature of 350 C (623 K) at $X = 0$ increases with increasing thermal conductivity of pine wood as k^2. This exposure time increase with increasing thermal conductivity of pine wood is due to decreasing temperature gradient at the surface of the semi-infinite solid.

Thermal conductivity of a pine log can change in radial direction, in perpendicular to grain direction and with the amount of its moisture content. Change in exposure time for pine wood's surface to reach 350 C divided by the change in pine wood's thermal conductivity versus pine wood's thermal conductivity is shown in Figure 16-7 for a total heat transfer coefficient of 15 $W/\left(m^2-K\right)$ and for other parameters are given above as constants.

Change in exposure time for pine wood's surface to reach 350 C divided by the change in pine wood's thermal conductivity has a linear relationship to pine wood's thermal conductivity as given in Equation 16-5.

$$\frac{\partial \theta}{\partial k} = 1.9 \frac{k}{\alpha h_{total}^{3}} \qquad 16\text{-}5$$

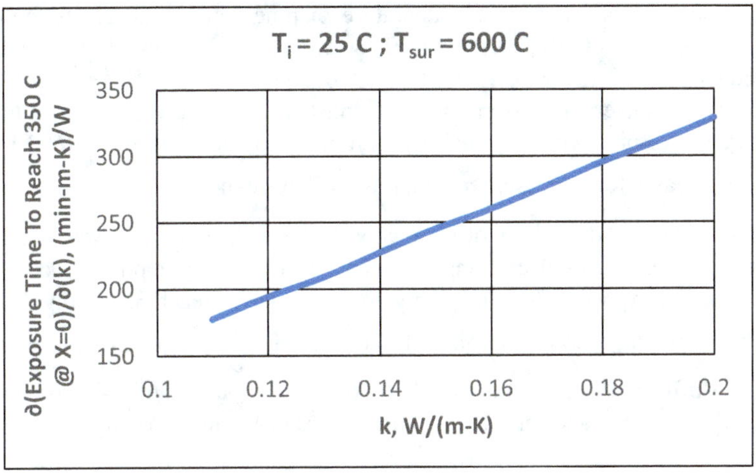

Figure 16-7 Change in exposure time for pine wood's surface to reach 350 C divided by the change in pine wood's thermal conductivity versus pine wood's thermal conductivity for a total heat transfer coefficient of $15\ W/\left(m^2-K\right)$

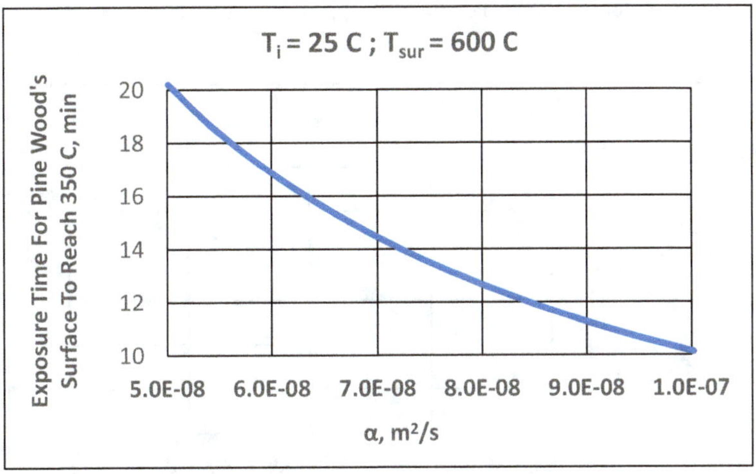

Figure 16-8 Exposure time for pine wood's surface to reach 350 C versus pine wood's thermal diffusivity for a total heat transfer coefficient of $15\ W/\left(m^2-K\right)$

Exposure times for surface temperature of pine wood to reach its ignition temperature of 350 C (623 K) at $X = 0$ for different α is shown in Figure 16-8 per Equation 16-3 for a total heat transfer coefficient of 15 $W/\left(m^2 - K\right)$ and for other parameters are given above as constants. Exposure times for surface temperature of pine wood to reach its ignition temperature of 350 C (623 K) at $X = 0$ decreases for increasing α values as $\frac{1}{\alpha}$. This exposure time decrease with increasing thermal diffusivity of pine wood is due to increasing conduction heat transfer rate along with decreasing volumetric heat capacity in pine wood.

Change in exposure time for pine wood's surface to reach 350 C divided by the change in pine wood's thermal diffusivity has a $-\frac{1}{\alpha^2}$ relation to pine wood's thermal diffusivity as shown in Equation 16-6 for a total heat transfer coefficient of 15 $W/\left(m^2 - K\right)$ and for other parameters are given above as constants.

$$\frac{\partial \theta}{\partial \alpha} = -0.95 \frac{k^2}{\alpha^2 h_{total}^3} \qquad \text{16-6}$$

Change in exposure time for pine wood's surface to reach 350 C divided by the change in pine wood's thermal diffusivity has a $\left(-\frac{1}{\alpha^2}\right)$ proportional relationship to pine wood's thermal diffusivity is shown in Figure 16-9. Absolute value of a change in exposure time for pine wood's surface to reach 350 C divided by the change in pine wood's thermal diffusivity approaches zero as α increases.

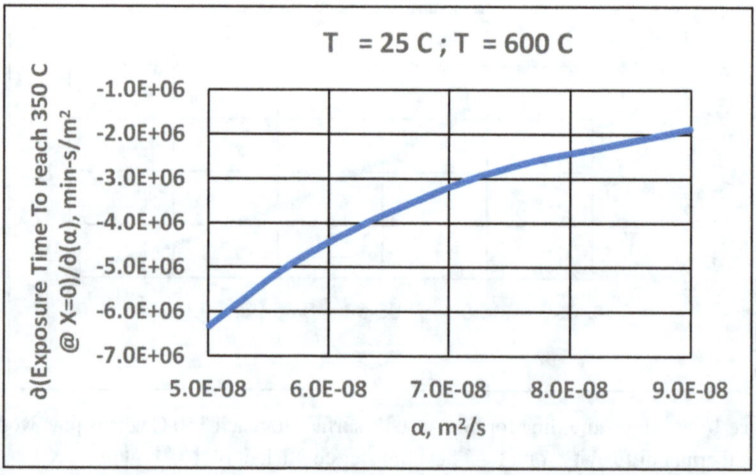

Figure 16-9 Change in exposure time for pine wood's surface to reach 350 C divided by the change in pine wood's thermal diffusivity versus pine wood's thermal diffusivity for a total heat transfer coefficient of 15 $W/\left(m^2 - K\right)$

17

Heat Transfer in A Semi-Infinite Stainless-Steel When Applying Constant Heat Flux On $\boldsymbol{X} = 0$ Surface

We will investigate in this chapter the heat transfer in a semi-infinite stainless-steel when its surface at $X = 0$ is suddenly exposed to a constant heat flux from ceramic heat treatment pads. One-dimensional transient conduction heat transfer in a semi-infinite solid which is suddenly exposed to a constant heat flux at $X = 0$ will be treated. There is suddenly a constant surface flux, $q_{X=0}$, which is applied at $X = 0$ surface as shown in Figure 17-1. Applicable partial differential equation and initial and boundary conditions are given in Equations 17-1.

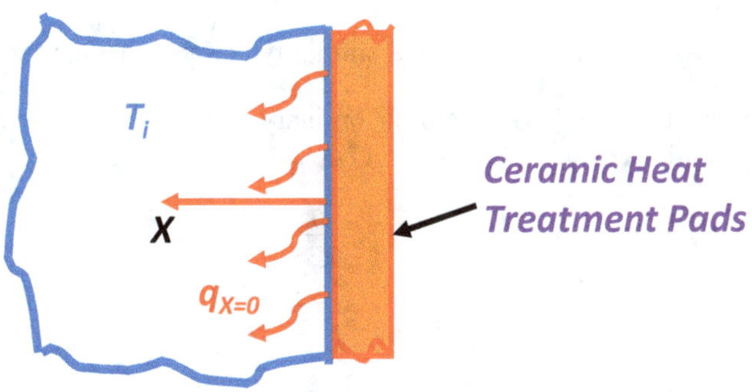

Figure 17-1 Semi-infinite solid which is initially at a temperature T_i and which is exposed suddenly to a constant surface heat flux, $q_{X=0}$, applied at $X = 0$

$$\frac{\partial^2 T}{\partial X^2} = \frac{1}{\alpha}\frac{\partial T}{\partial \theta} \qquad \text{17-1a}$$

$$T(X,0) = T_i \qquad \text{17-1b}$$

$$(X \to \infty, \theta) = T_i \qquad \text{17-1c}$$

$$\left(-k\frac{\partial T}{\partial X}\right)_{X=0} = q_{X=0} \qquad \text{17-1d}$$

$T(X,\theta)$ is the temperature of the semi-infinite solid in K or in C at X location in meters from its surface and at a time θ in seconds. k is semi-infinite solid's thermal conductivity in W/(m-K) and α is semi-infinite solid's thermal diffusivity in m²/s. $q_{X=0}$ is constant surface flux applied suddenly at $X = 0$ surface in W/m². We can introduce a new dependent variable, $q(X,\theta) = -k\frac{\partial T}{\partial X}$, which is the local heat flux within the semi-infinite solid body known as the Fourier's law. After differentiating both sides of Equation 17-1a with respect to X and then applying Fourier's law, we get Equations 17-2 where $q(X,\theta)$ is the new dependent variable with its initial and boundary conditions.

$$\frac{\partial^2 q}{\partial X^2} = \frac{1}{\alpha}\frac{\partial q}{\partial \theta} \qquad \text{17-2a}$$

$$q(X,0) = 0 \qquad \text{17-2b}$$

$$q(X \to \infty, \theta) = 0 \qquad \text{17-2c}$$

$$q(0, \theta > 0) = q_{X=0} \qquad \text{17-2d}$$

Let $\xi = \left(\frac{X}{2\sqrt{\alpha\theta}}\right)$ be the dimensionless similarity variable that converts the partial differential Equation 17-2a into an ordinary differential equation with its boundary conditions given in Equations 17-3.

$$\frac{\partial^2 q}{\partial \xi^2} = -2\xi\frac{\partial q}{\partial \xi} \qquad \text{17-3a}$$

$$q(\xi = 0) = q_{X=0} \qquad \text{17-3b}$$

$$q(\xi \to \infty) = 0 \qquad \text{17-3c}$$

The solution to Equations 17-3 is presented in Equation 17-4 as the heat flux, $q(X,\theta)$ in a semi-infinite solid suddenly exposed to a constant heat flux $q_{X=0}$.

$$q(X,\theta) = q_{X=0}erfc\left(\frac{X}{2\sqrt{\alpha\theta}}\right) \quad 17\text{-}4$$

where $erfc(\xi) = 1 - \text{erf}(\xi)$ and it is called the complementary error function. erf (ξ), the error function, was presented in Chapter 15, Equation 15-4.

We can now apply Fourier's law to Equation 17-4 to obtain the temperature distribution in the semi-infinite solid with the following integral given in Equation 17-5.

$$\int_{T_i}^{T(X,\theta)} dT = -\frac{q_{X=0}}{k}\int_{\infty}^{X} erfc\left(\frac{X}{2\sqrt{\alpha\theta}}\right)dX = -\frac{2q_{X=0}\sqrt{\alpha\theta}}{k}\int_{\infty}^{\xi} erfc(\xi)\,d\xi \quad 17\text{-}5$$

The integral of the complementary error function is provided in Equation 17-6.

$$\int erfc(\xi)\,d\xi = \xi \times erfc(\xi) - \frac{\exp(-\xi^2)}{\sqrt{\pi}} + C \quad 17\text{-}6$$

Combining Equation 17-5 and 17-6 provides us with the transient temperature distribution in a semi-infinite solid which is suddenly exposed to a constant heat flux $q_{X=0}$ and which satisfies Equations 17-1 is shown in Equation 17-7.

$$T(X,\theta) - T_i = \frac{q_{X=0}}{k}\left[\left(2\sqrt{\frac{\alpha\theta}{\pi}}\right)e^{\left(-\frac{X^2}{4\alpha\theta}\right)} - (X)erfc\left(\frac{X}{2\sqrt{\alpha\theta}}\right)\right] \quad 17\text{-}7$$

Constant properties that are used in following calculations for a large stainless-steel suddenly exposed to a constant heat flux from ceramic heat treatment pads are provided below.

$$T_i = 25C; \qquad q_{X=0} = 50{,}000\ W/m^2$$

$$\alpha = 3.5\times10^{-6}\frac{m^2}{s}; \qquad k = 13.5\frac{W}{m-C}$$

Figure 17-2 shows the temperature versus time curves at different *X* locations in a semi-infinite solid suddenly exposed to a constant heat flux of 50,000 W/m^2 from ceramic heat treatment pads. These temperature versus time curves are obtained from Equation 17-7. At $X = 0$ surface, the temperature increases from 25 C to 688 C in two hours. As we go deeper into the semi-infinite solid, temperature increase with respect to time decreases. For example, at $X = 0.4$ *m*, the temperature only increases from 25 C to 50 C in two hours.

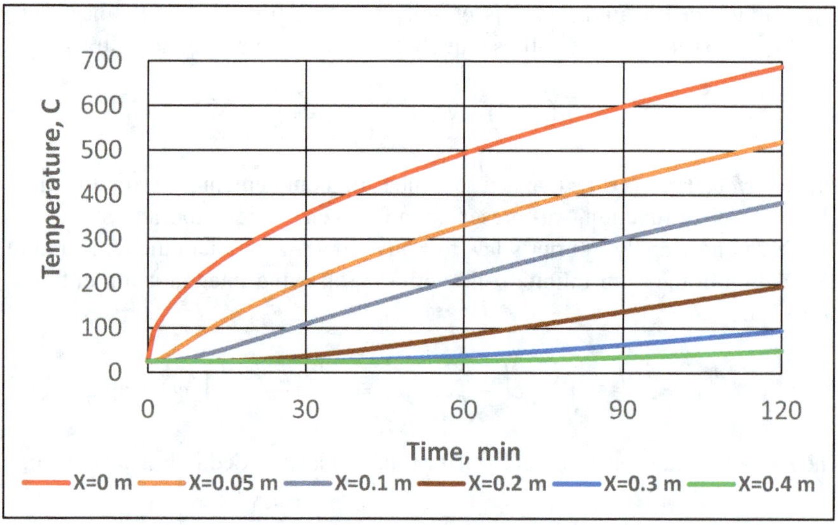

Figure 17-2 Temperature versus time at different *X* locations in a semi-infinite solid suddenly exposed to a constant heat flux of 50,000 W/m^2 from ceramic heat treatment pads using Equation 17-6

Figure 17-3 shows temperature versus *X* location at different times in a semi-infinite solid suddenly exposed to a constant heat flux of 50,000 W/m^2 from ceramic heat treatment pads. These temperature versus *X* location curves at different times are also obtained from Equation 17-7. As time increases, most affected location is at the surface of the semi-infinite solid. As time increases, temperature increase decreases as we go deeper into the semi-infinite solid.

Figure 17-4 shows heat fluxes within the semi-infinite solid versus *X* location at different times. These curves are slopes of the curves in Figure 17-3. They are obtained using Equation 17-4. As time increases heat fluxes within the semi-infinite solid increase at all *X* locations. The highest heat flux is at the surface of the semi-infinite solid which is at 50,000 W/m^2 in this case.

Now, let us investigate sensitivities of temperatures at the surface of the semi-infinite solid to changes in $q_{X=0}$, α, and k.

At $X = 0$, Equation 17-7 becomes as shown in Equation 17-8.

$$T(0,\theta) - T_i = \frac{2q_{X=0}}{k}\sqrt{\frac{\alpha\theta}{\pi}} \qquad \text{17-8}$$

Figure 17-5 shows the temperature increase at $X = 0$ with increasing $q_{X=0}$ at different times. The heat flux at $X = 0$ is varied $50,000 \pm 10\%\ W/m^2$ for present sensitivity calculations. At $X = 0$, temperature increases linearly with increasing surface heat flux which can be deducted from Equation 17-8. In Figure 17-5, temperature versus heat flux at $X = 0$ lines move upward as time increases as the square-root of θ.

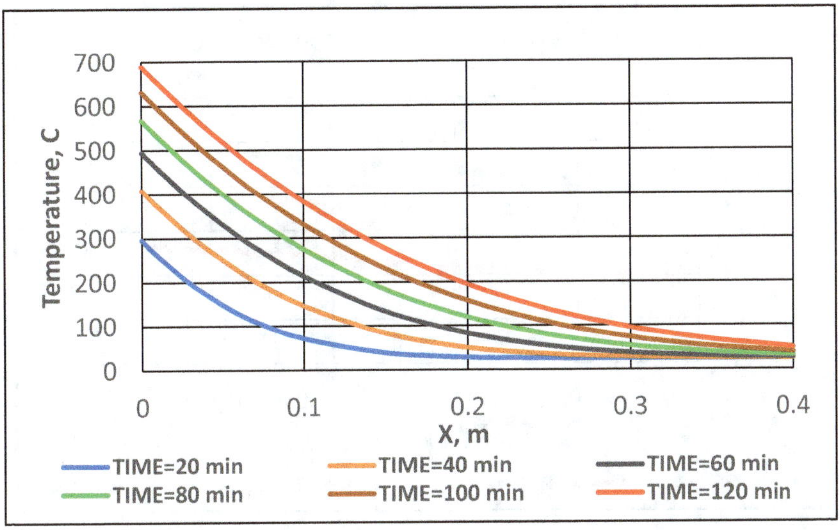

Figure 17-3 Temperature versus X location at different times in a semi-infinite solid suddenly exposed to a constant heat flux of 50,000 W/m^2 from ceramic heat treatment pads using Equation 17-6

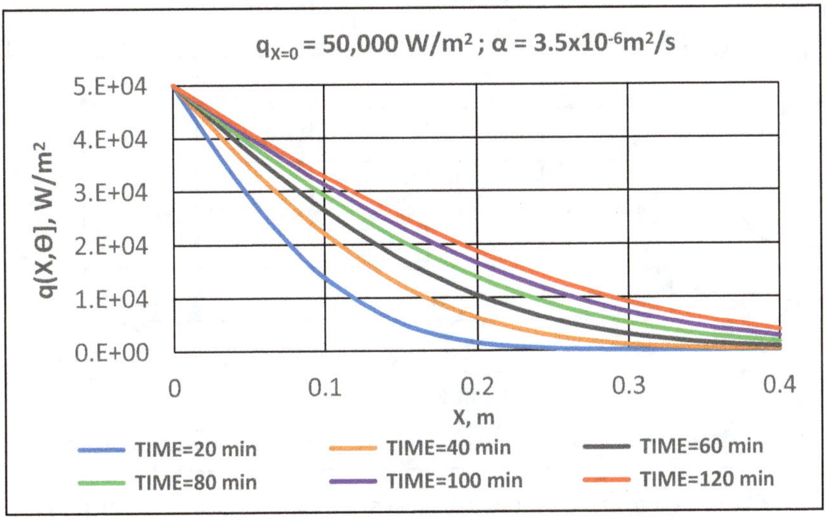

Figure 17-4 Heat flux within the semi-infinite solid versus X location at different times from Equation 17-3

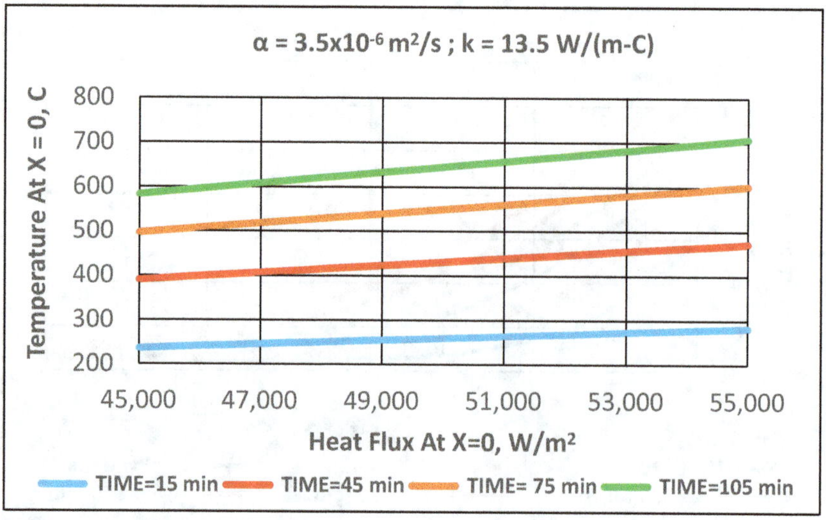

Figure 17-5 Temperature increase at $X = 0$ with increasing $q_{X=0}$ at different times

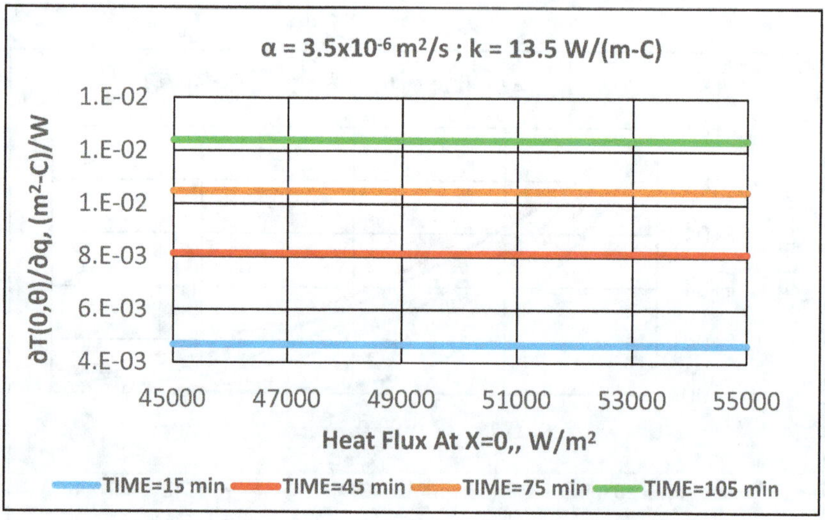

Figure 17-6 Sensitivity of semi-infinite solid's surface temperature to changing $q_{X=0}$ versus $q_{X=0}$ at different times as given in Equation 17-8

Sensitivity of semi-infinite solid's surface temperature to changing $q_{X=0}$ is shown in Equation 17-9 while all other parameters are constants and it is presented in Figure 17-6 for different times.

$$\frac{\partial T(0,\theta)}{\partial q_{X=0}} = \frac{2}{k}\sqrt{\frac{\alpha\theta}{\pi}} \qquad \text{17-9}$$

Sensitivity of semi-infinite solid's surface temperature to changing $q_{X=0}$ is a constant for constant values of θ, α, and k.

Figure 17-7 shows the temperature decrease at $X = 0$ with increasing thermal conductivity, k, at different times. The semi-infinite solid's thermal conductivity is varied $13.5 \pm 1.5\ W/(m-C)$ for present sensitivity calculations. As semi-infinite solid's conduction heat transfer rate increases, $T(0,\theta)$ decreases as $\frac{1}{k}$. See Equation 17-8. The sensitivity of $T(0,\theta)$ to changing thermal conductivity is given in Equation 17-10 while all other parameters are constants.

$$\frac{\partial T(0,\theta)}{\partial k} = -\frac{2q_{X=0}}{k^2}\sqrt{\frac{\alpha\theta}{\pi}} \qquad \text{17-10}$$

Absolute value of sensitivity of $T(0,\theta)$ to increasing thermal conductivity decreases as depicted in Figure 17-8. As time increases, the absolute value of sensitivity of $T(0,\theta)$ to increasing conduction heat transfer rate becomes more prominent.

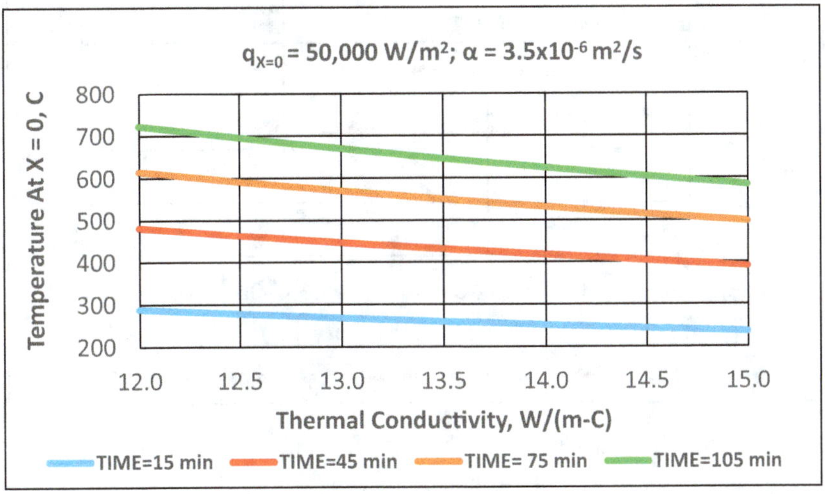

Figure 17-7 Temperature decrease at $X = 0$ with increasing thermal conductivity, k, at different times

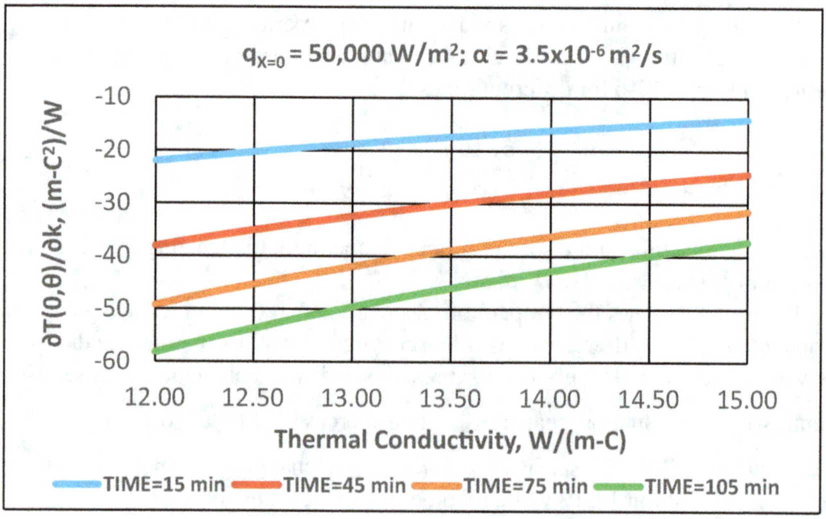

Figure 17-8 Sensitivity of $T(0,\theta)$ to changing thermal conductivity as given in Equation 17-9

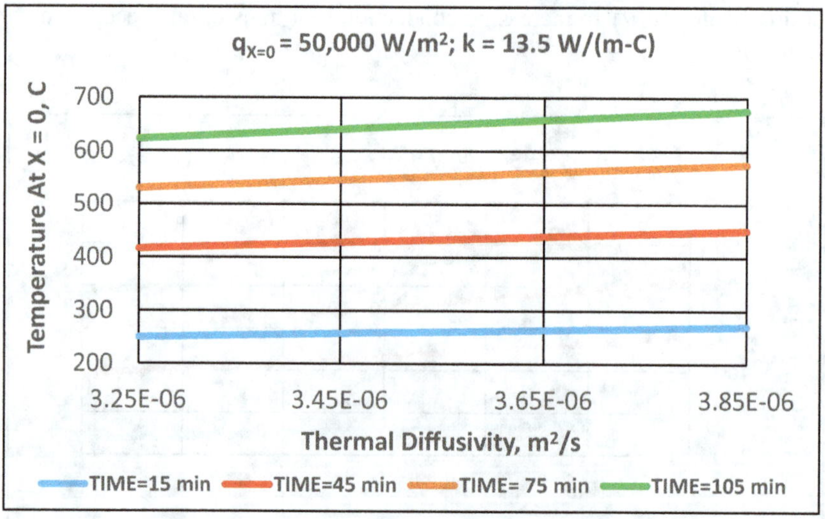

Figure 17-9 Temperature increase at $X = 0$ with increasing thermal diffusivity, α, at different times

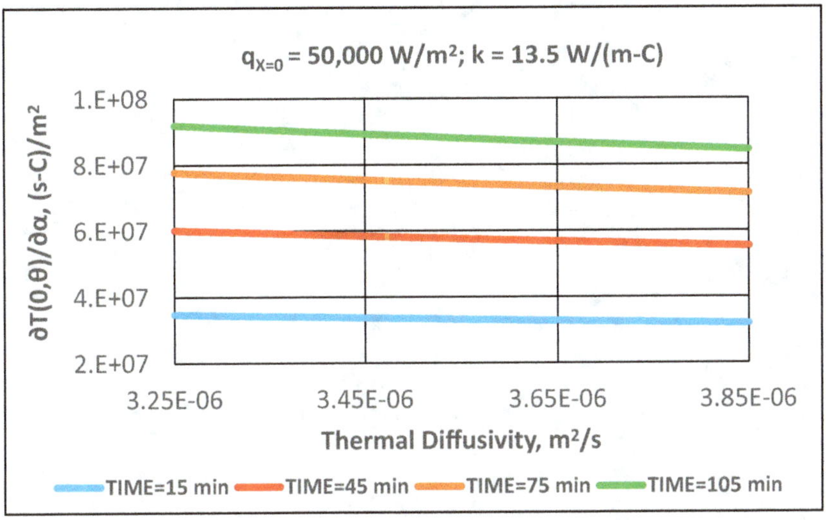

Figure 17-10 Sensitivity of $T(0,\theta)$ to increasing thermal diffusivity as given in Equation 17-10

Figure 17-9 shows the temperature increase at $X = 0$ with increasing thermal diffusivity, α, at different times. The semi-infinite solid's thermal diffusivity is varied from $3.25\times10^{-6}\ m^2/s$ to $3.85\times10^{-6}\ m^2/s$ for present sensitivity calculations. As semi-infinite solid's volumetric heat capacity or thermal diffusivity increases, $T(0,\theta)$ increases as $\sqrt{\alpha}$. See Equation 17-9. Sensitivity of $T(0,\theta)$ to changing thermal diffusivity is given in Equation 17-11 while all other parameters are constants.

$$\frac{\partial T(0,\theta)}{\partial \alpha} = \frac{q_{X=0}}{k}\sqrt{\frac{\theta}{\pi\alpha}} \qquad 17\text{-}11$$

Sensitivity of $T(0,\theta)$ to increasing thermal diffusivity decreases at a given time as depicted in Figure 17-10. Change in semi-infinite solid's volumetric heat capacity is less effective at low semi-infinite solid temperatures as expected.

18

Solidification of Molten Surface Coating on a Semi-Infinite Solid Substrate

In this chapter, we will investigate surface coating by thermal spraying of molten material on a semi-infinite solid substrate. Wire or powder flame spraying on the surface of a solid substrate is a very reliable and cost-effective process used for corrosion, erosion and wear resistance in automotive and aerospace industries and on bridges, ships and other structures. Aluminum oxide and zinc alloys are most widely used molten materials for flame spraying on iron and steel.

There is a constant surface temperature applied to $X = 0$ surface of a semi-infinite solid substrate from sprayed molten coating material which has a solidified thickness, *th*, in the order of millimeters, see Figure 18-1, and it has a much higher initial temperature than semi-infinite solid's initial

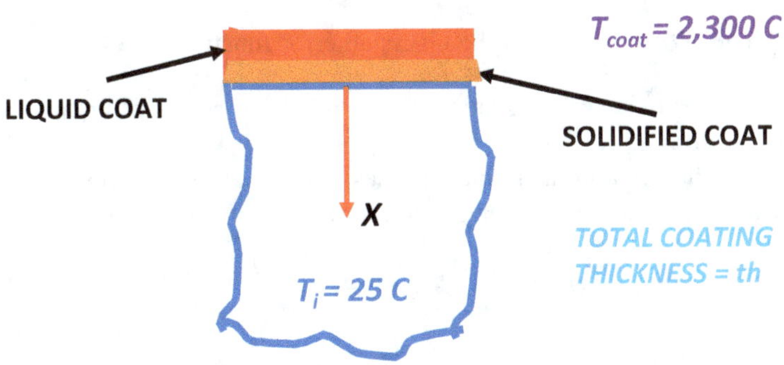

Figure 18-1 Semi-infinite solid is initially at a temperature T_i. It is suddenly exposed to a constant surface temperature $\boldsymbol{T_{coat}}$ where $T_i << T_{coat}$. Molten coating material covers the surface of the semi-infinite solid and solidifies fast

temperature, i.e., 2,300 C versus 25 C. Following partial differential equation and initial and boundary conditions apply to a semi-infinite solid substrate which is suddenly exposed to molten coating material on its surface are shown in Equations 18-1.

$$\frac{\partial^2 T}{\partial X^2} = \frac{1}{\alpha}\frac{\partial T}{\partial \theta} \qquad \text{18-1a}$$

$$T(X,0) = T_i \qquad \text{18-1b}$$

$$T(X \rightarrow \infty, \theta) = T_i \qquad \text{18-1c}$$

$$T(0,\theta) = T_{coat} \qquad \text{18-1d}$$

$T(X,\theta)$ is the temperature in the semi-infinite solid at the X location in meters from its surface and at a time θ in seconds. The temperature distribution in a semi-infinite solid that satisfies Equations 18-1 is shown in Equation 18-2. The heat transfer at the surface of the semi-infinite solid, $X = 0$, is given in Equation 18-3.

$$\frac{T(X,\theta) - T_{coat}}{T_i - T_{coat}} = erf\left(\frac{X}{2\sqrt{\alpha_s \theta}}\right) \qquad \text{18-2}$$

$$Q(0,\theta) = \frac{k_s (T_{coat} - T_i)}{\sqrt{\pi \alpha_s \theta}} \qquad \text{18-3}$$

where k_s is semi-infinite solid's thermal conductivity in W/(m-K) and α_s is semi-infinite solid's thermal diffusivity in m²/s. $\left(\frac{X}{2\sqrt{\alpha_s \theta}}\right)$ is the dimensionless similarity variable that converts the partial differential Equation 18-1a into an ordinary one. $erf\left(\frac{X}{2\sqrt{\alpha_s \theta}}\right)$ is the Gaussian error function. See Chapter 15 for details of the Gaussian error function. For detailed derivations of Equations 18-2 and 18-3, see Ref. [10].

As the molten coating material solidifies, it releases its latent heat of fusion, ΔH_{coat}, in J/kg to the semi-infinite solid's surface as given in Equation 18-4.

$$\rho_{coat} \Delta H_{coat} \frac{d(th)}{d\theta} = Q(0,\theta) = \frac{k_s (T_{coat} - T_i)}{\sqrt{\pi \alpha_s \theta}} \qquad \text{18-4}$$

where ρ_{coat} is the density of coating material in kg/m³, *th*, is the solidified coating material's thickness in meters, θ, is time in seconds, k_s, is the thermal conductivity of semi-infinite solid in W/(m-C), α_s, is the thermal diffusivity of semi-infinite solid in m^2/s, T_i, is the initial temperature of the semi-infinite solid in C, T_{coat} is the temperature of the coating material both in liquid and in solid phases during solidification in C. By integrating Equation 18-4, we can obtain the complete solidification time, θ_{cs} , for the molten coating material as shown in Equation 18-5.

$$\theta_{cs} = \pi\alpha_s \left[\frac{th \times \rho_{coat} \Delta H_{coat}}{2k_s \left(T_{coat} - T_i\right)} \right]^2 \tag{18-5}$$

Now, let us analyze effect of each variable on complete solidification time of molten surface coat using Equation 18-5. Let us use the following nominal values for each variable. Then, let us investigate the sensitivity of $\pm 10\%$ variation in each variable on complete solidification time, θ_{cs} .

$\alpha_s = 1.90 \times 10^{-5} \ m^2/s$; $\qquad k_s = 50 \ W/(m-C)$

$\rho_{coat} = 4{,}000 \ kg/m^3$; $\qquad \Delta H_{coat} = 1.26 \times 10^6 \ J/kg$

$T_{coat} = 2{,}300 \ C$; $\qquad T_i = 25 \ C$

Complete solidification times of molten surface coat versus solid substrate's thermal diffusivity for four different surface coat thicknesses are shown in Figure 18-2. For surface coat thicknesses of 1 mm and less, $\pm 10\%$ variation in thermal diffusivity of semi-infinite solid does not affect complete solidification times. As surface coat thickness increases over 1 mm, so does the effect of varying thermal diffusivity of semi-infinite solid on complete solidification times. Complete solidification times increases linearly with respect to increasing thermal diffusivity of semi-infinite solid. Change in complete solidification time of molten surface coat to change in solid substrate's thermal diffusivity versus solid substrate's thermal diffusivity for four different surface coat thicknesses are shown in Figure 18-3.

Sensitivity of complete solidification time to changing solid substrate's thermal diffusivity is constant and the value of this constant increases as the square of the increasing surface coat thickness, *th*.

Figure 18-4 shows the complete solidification times versus thickness of surface coating for three different solid substrate's thermal diffusivities. We can see the effect of changing thermal diffusivity with increasing *th*.

Complete solidification time of molten surface coat versus thickness of surface coat is shown in Figure 18-5. There is a parabolic relationship between θ_{cs} and *th*. Complete solidification time of molten surface coat increases as $(th)^2$. Change in complete solidification time of molten surface coat to change in thickness of surface coat versus thickness of surface coat is shown in Figure 18-6. Sensitivity of complete solidification time to changing thickness of surface coat is a linear relationship. This sensitivity increases linearly with increasing *th* as shown in Figure 18-6.

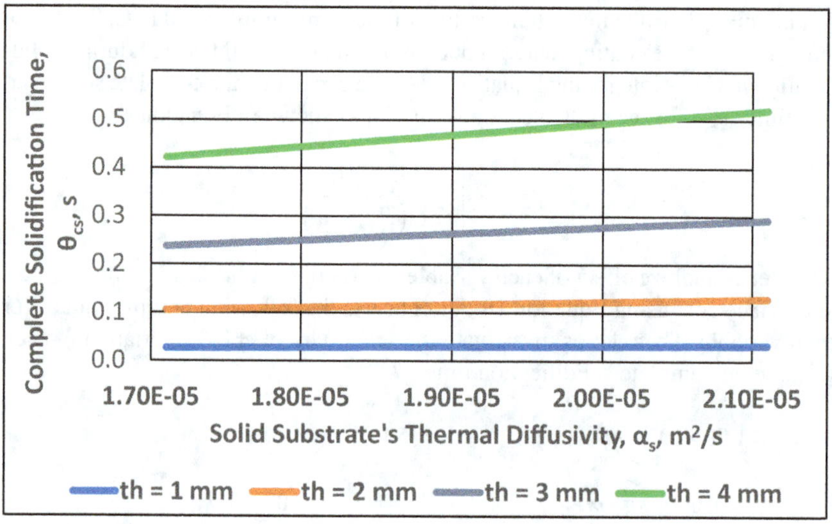

Figure 18-2 Complete solidification time of molten surface coat versus solid substrate's thermal diffusivity for four different surface coat thicknesses

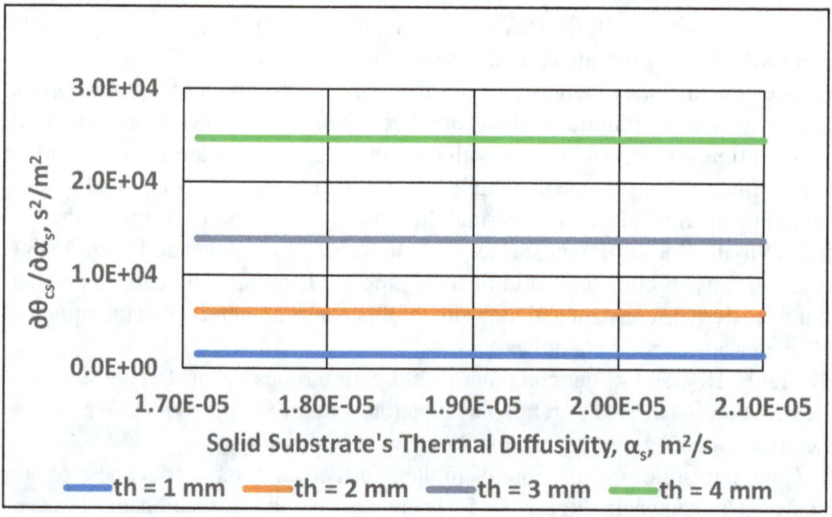

Figure 18-3 Change in complete solidification time of molten surface coat to change in solid substrate's thermal diffusivity versus solid substrate's thermal diffusivity for four different surface coat thicknesses

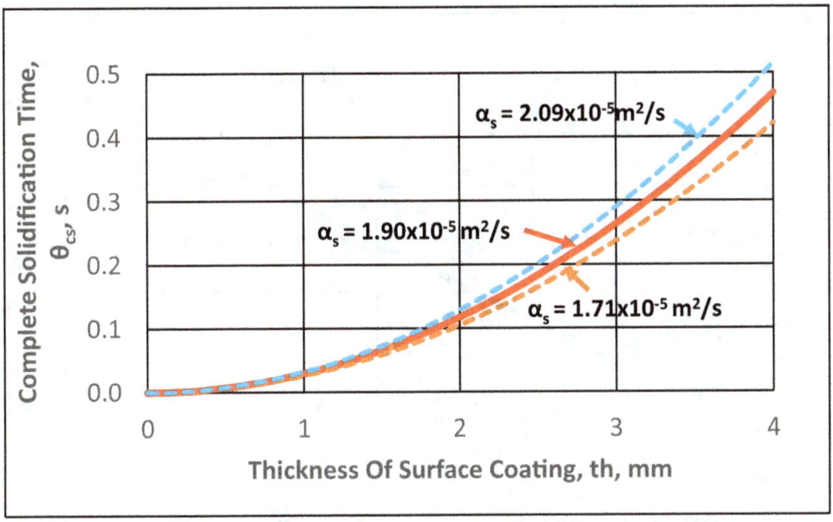

Figure 18-4 Complete solidification time of molten surface coat versus thickness of surface coating for nominal and ±10% of nominal solid substrate's thermal diffusivities

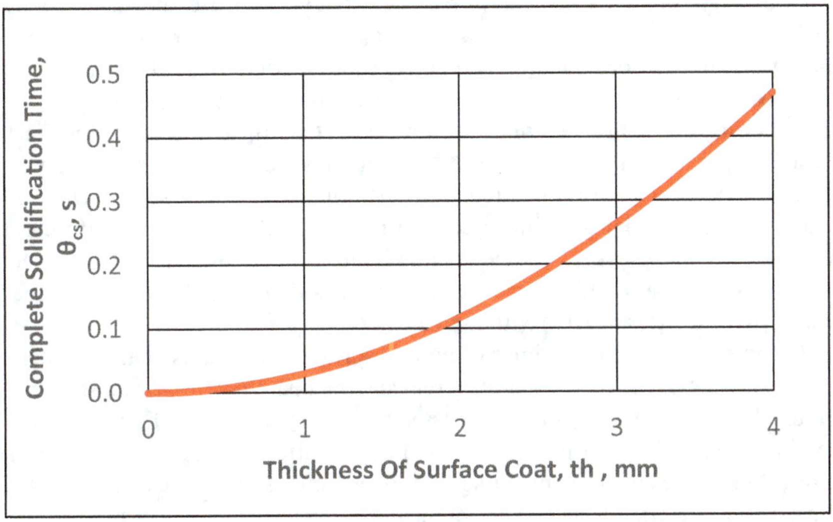

Figure 18-5 Complete solidification time of molten surface coat versus thickness of surface coat

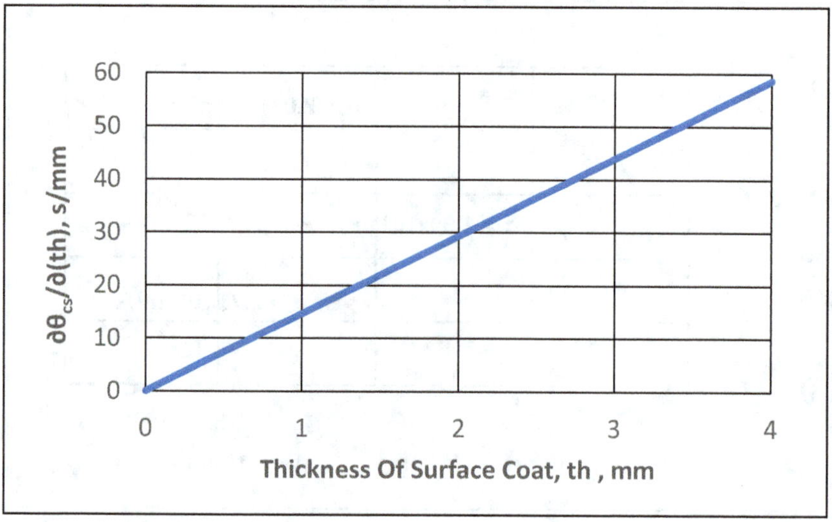

Figure 18-6 Change in complete solidification time of molten surface coat to change in thickness of surface coat versus thickness of surface coat

Complete solidification times of molten surface coat versus coating material's density for four different surface coat thicknesses are shown in Figure 18-7. For surface coat thicknesses of 1 mm and less, ±10% variation in coating material's density does not affect complete solidification times. As surface coat thickness increases over 1 mm, so does the effect of varying coating material's density on complete solidification times. Complete solidification times increases parabolically with respect to increasing coating material's density. Change in complete solidification time of molten surface coat to change in coating material's density versus coating material's density for four different surface coat thicknesses are shown in Figure 18-8. Sensitivity of complete solidification time to changing coating material's density is linear. This linear sensitivity increases as $(th)^2$ with increasing *th*.

Complete solidification times of molten surface coat versus coating material's latent heat of fusion for four different surface coat thicknesses are shown in Figure 18-9. For surface coat thicknesses of 1 mm and less, ±10% variation in coating material's latent heat of fusion does not affect complete solidification times. As surface coat thickness increases over 1 mm, so does the effect of varying coating material's latent heat of fusion on complete solidification times. Complete solidification times increases parabolically with respect to increasing coating material's latent heat of fusion. Change in complete solidification time of molten surface coat to change in coating material's latent heat of fusion versus coating material's latent heat of fusion for four different surface coat thicknesses are shown in Figure 18-10. Sensitivity of complete solidification

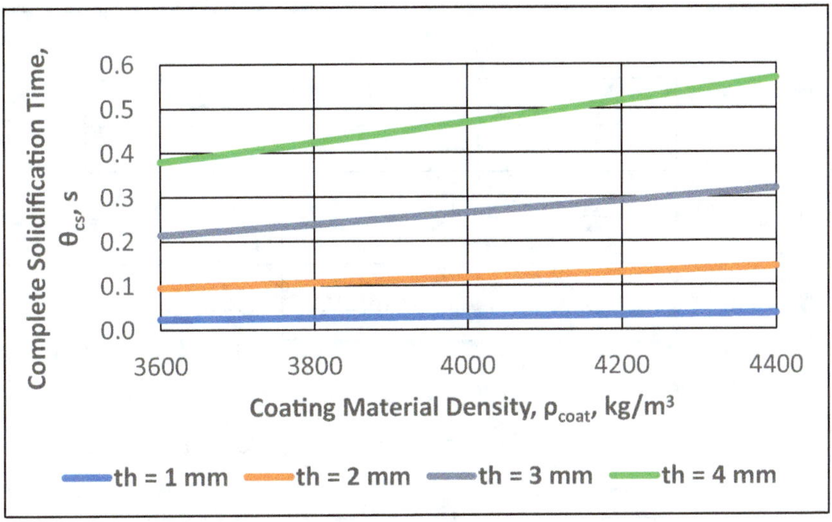

Figure 18-7 Complete solidification time of molten surface coat versus coating material's density for four different surface coat thicknesses

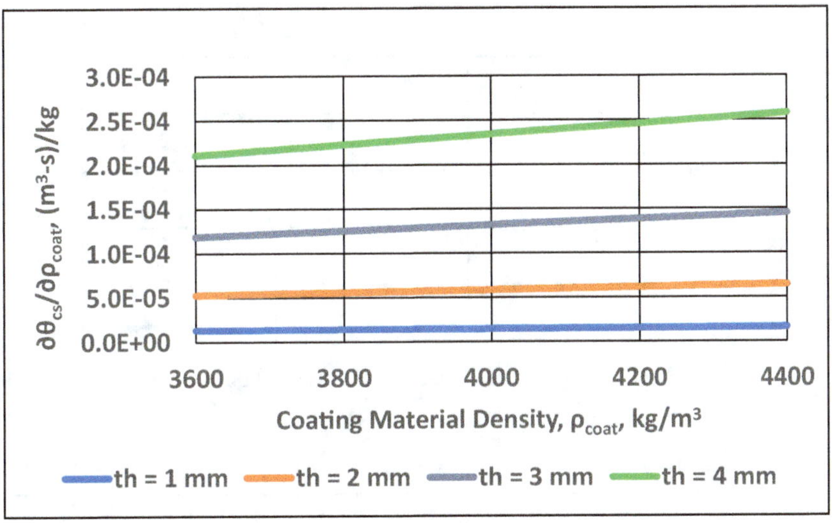

Figure 18-8 Change in complete solidification time of molten surface coat to change in coating material's density versus coating material's density

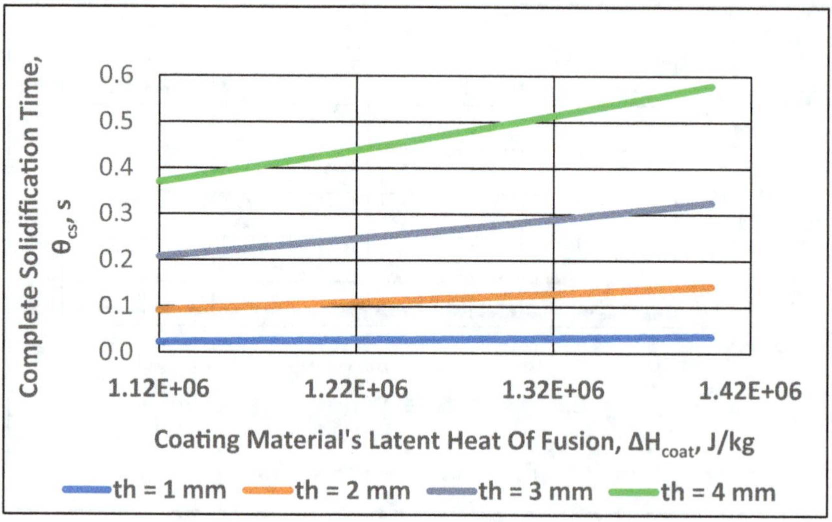

Figure 18-9 Complete solidification times of molten surface coat versus coating material's latent heat of fusion for four different surface coat thicknesses

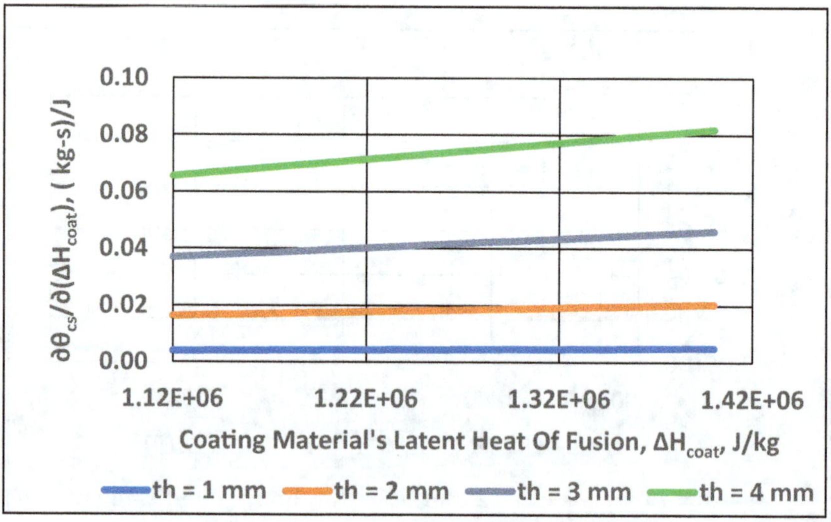

Figure 18-10 Change in complete solidification time of molten surface coat to change in coating material's latent heat of fusion versus coating material's latent heat of fusion for four different surface coat thicknesses

time to changing coating material's latent heat of fusion is linear. This linear sensitivity increases as $(th)^2$ with increasing *th*.

Complete solidification times of molten surface coat versus solid substrate's thermal conductivity for four different surface coat thicknesses are shown in Figure 18-11. For surface coat thicknesses of 1 mm and less, $\pm 10\%$ variation in thermal conductivity of semi-infinite solid does not affect complete solidification times. As surface coat thickness increases over 1 mm, so does the effect of varying thermal conductivity of semi-infinite solid on complete solidification times. Complete solidification times decreases as $\left(\frac{1}{k_s^2}\right)$ with respect to increasing thermal conductivity of semi-infinite solid. Change in complete solidification time of molten surface coat to change in solid substrate's thermal conductivity versus solid substrate's thermal conductivity for four different surface coat thicknesses are shown in Figure 18-12. Sensitivity of complete solidification time to change in solid substrate's thermal conductivity is $\left(-\frac{1}{k_s^3}\right)$ and the absolute value of this sensitivity increases as the square of the increasing surface coat thickness.

Complete solidification times of molten surface coat versus $(T_{coat} - T_i)$ for four different surface coat thicknesses are shown in Figure 18-13. For surface coat thicknesses of 1 mm and less, $\pm 10\%$ variation in $(T_{coat} - T_i)$ does affect complete solidification times a little. As surface coat thickness increases over 1 mm, so does the effect of varying $(T_{coat} - T_i)$ on complete solidification times.

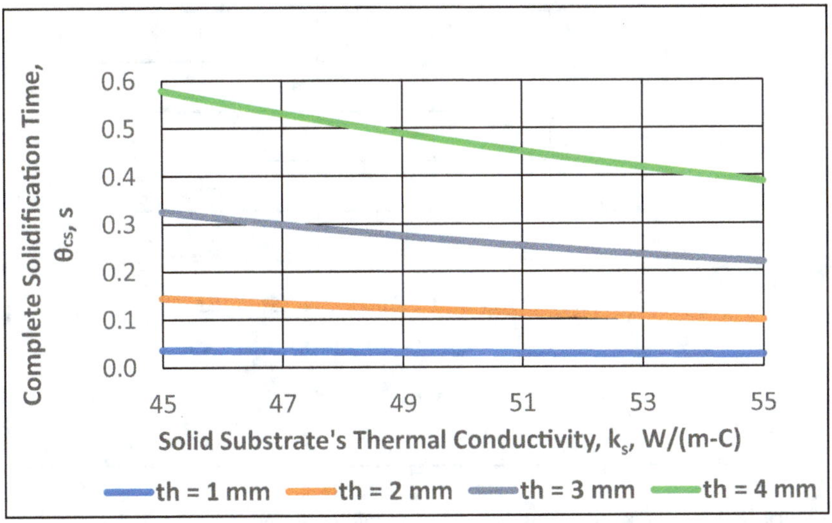

Figure 18-11 Complete solidification time of molten surface coat versus solid substrate's thermal conductivity for four different surface coat thicknesses

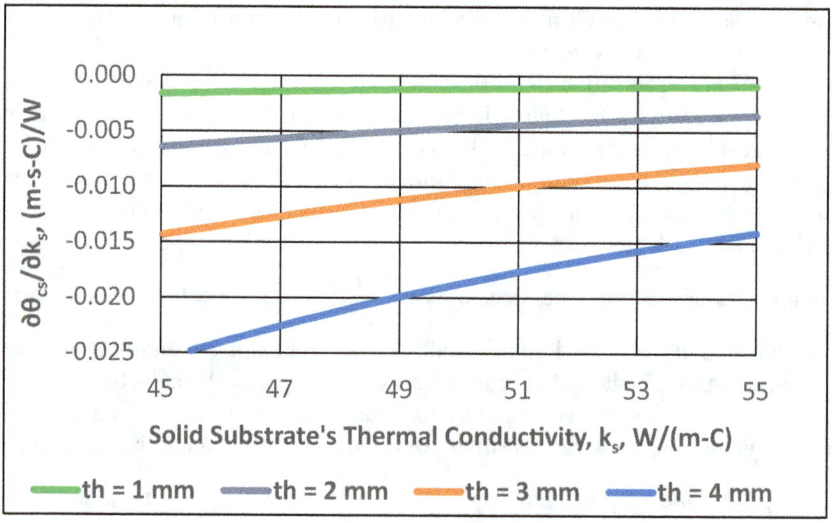

Figure 18-12 Change in complete solidification time of molten surface coat to change in solid substrate's thermal conductivity versus solid substrate's thermal conductivity for four different surface coat thicknesses

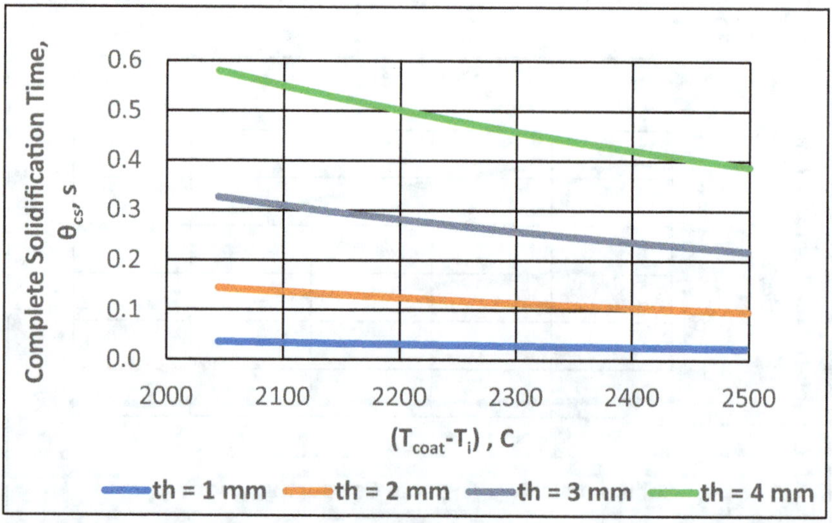

Figure 18-13 Complete solidification time of molten surface coat versus $(T_{coat} - T_i)$ for four different surface coat thicknesses

Complete solidification times decreases as $\left[\frac{1}{(T_{coat}-T_i)^2}\right]$ with respect to increasing $(T_{coat}-T_i)$. Change in complete solidification time of molten surface coat to change in $(T_{coat}-T_i)$ versus $(T_{coat}-T_i)$ for four different surface coat thicknesses are shown in Figure 18-14. Sensitivity of complete solidification time to changing $(T_{coat}-T_i)$ is $\left(-\frac{1}{(T_{coat}-T_i)^3}\right)$ and the absolute value of this sensitivity increases as the square of the increasing surface coat thickness.

Now, let us use Equation 18-2 to determine the temperature distributions in the semi-infinite solid, carbon steel in this case, after suddenly the molten surface coat is applied at 2,300 C. Spatial temperature distributions in the semi-infinite carbon steel are shown in Figure 18-15 for four different times. Heat penetrates up to 1.5 cm after 1 second and up to 4.5 cm after 10 seconds. Slopes of these temperature profiles give us the amount of conduction heat transferring in the semi-infinite solid at a given X location and at a time θ. Conduction heat transfer profiles are shown in Figure 18-17.

We can also investigate temperature profiles within the semi-infinite solid as a function of time for a given X location as shown in Figure. 18-16. At $X = 0.5\ cm$, the temperature rises fast and approaches 2,000 C in about 10 seconds. As we go deeper into the semi-infinite carbon steel, temperature increase

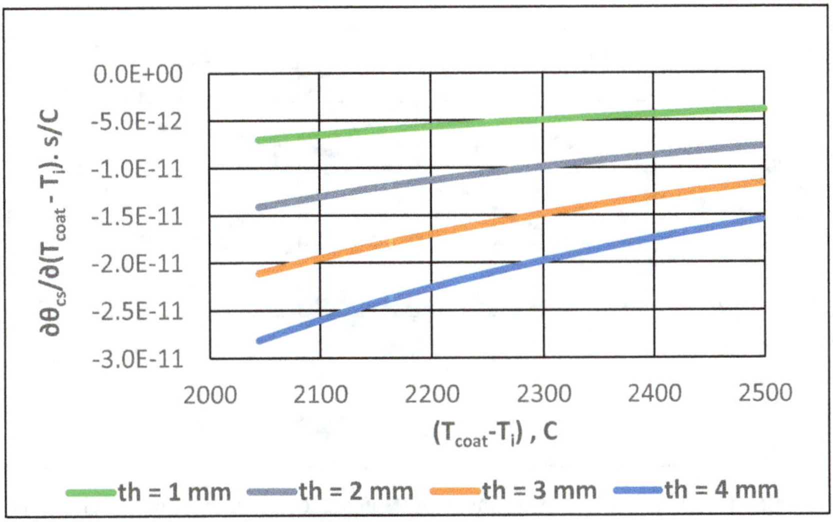

Figure 18-14 Change in complete solidification time of molten surface coat to change in $(T_{coat}-T_i)$ versus $(T_{coat}-T_i)$ for four different surface coat thicknesses

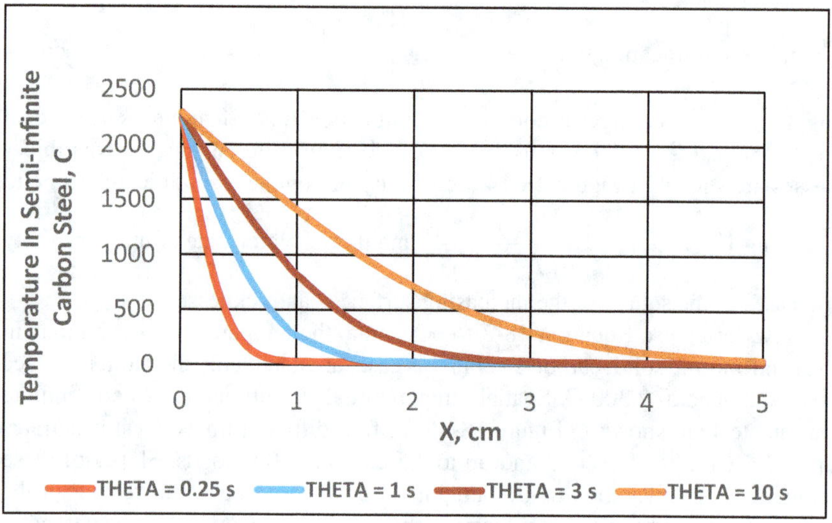

Figure 18-15 Spatial temperature distributions in the semi-infinite carbon steel for four different times

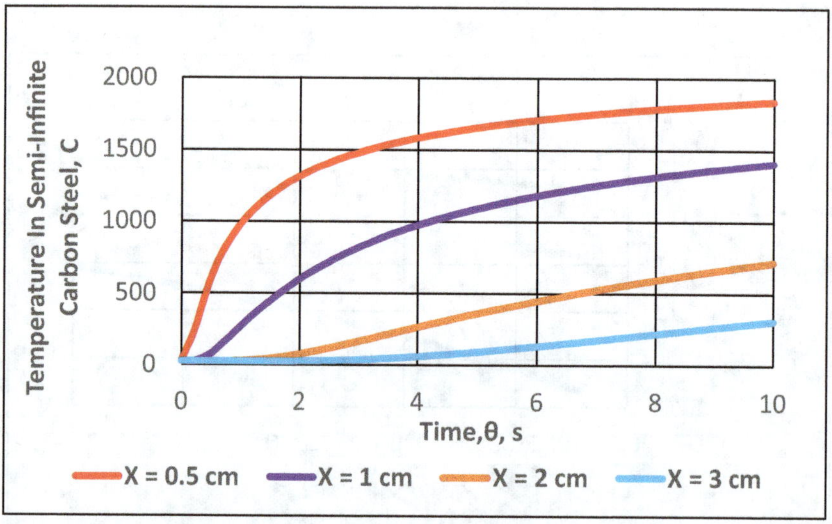

Figure 18-16 Temperature versus time profiles in the semi-infinite carbon steel for four different X locations

slows substantially because incoming energy from the molten surface coat is distributed into a very high volume.

Spatial conduction heat transfer distributions in the semi-infinite carbon steel can be obtained from the slopes of temperature profiles in Figure 18-15. A more exact way is to determine spatial conduction heat transfer distributions from the derivative of Equation 18-2 by applying Fourier's law, $Q(X,\theta) = -k\frac{\partial T(X,\theta)}{\partial X}$ in W/m^2 as shown in Equations 18-6.

$$Q(X,\theta) = -k\frac{\partial T(X,\theta)}{\partial X} = k(T_{coat} - T_i)\frac{d\left(erf\left(\frac{X}{2\sqrt{\alpha_s\theta}}\right)\right)}{d\left(\frac{X}{2\sqrt{\alpha_s\theta}}\right)} \times \frac{d\left(\frac{X}{2\sqrt{\alpha_s\theta}}\right)}{dX} \qquad 18\text{-}6a$$

$$Q(X,\theta) = \left[\frac{k(T_{coat} - T_i)}{\sqrt{\pi\alpha_s\theta}}\right]\exp\left(-\frac{X^2}{4\alpha_s\theta}\right) \qquad 18\text{-}6b$$

Conduction heat transfer within the semi-infinite solid decreases fast as time increases due to its very large volume. Decrease in conduction heat transfer at $X = 0$ is an order of magnitude less when $\theta = 10\ s$ than when $\theta = 0.25\ s$.

Spatial conduction heat transfer distributions in the semi-infinite carbon steel for four different times are shown in Figure 18-17.

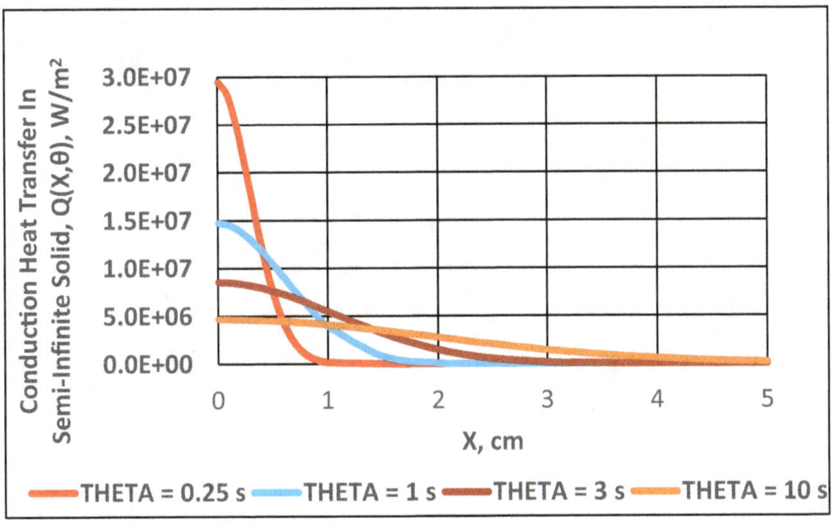

Figure 18-17 Spatial conduction heat transfer distributions in the semi-infinite carbon steel for four different times

19

Large Plane Concrete Wall Subjected Suddenly to Convection Heat Transfer

In this chapter, we will investigate transient temperature distribution in a large plane concrete wall which is subjected suddenly to convection heat transfer on both surfaces and the initial plane wall temperature, T_{w-ini}, does not equal to environment's temperature, T_{env}, or $T_{w-ini} \neq T_{env}$, as shown in Figure 19-1. There are several ways we can approach to solve transient temperature distributions in a large plane wall depending upon values of two important non-dimensional numbers. First non-dimensional number is the Biot number given in Equation 19-1 which is the ratio of the conduction heat transfer resistance within the plane wall to convection heat transfer resistance at two surfaces of the plane wall.

where L is half of the plane wall thickness in meters which is its characteristic length, k is the constant thermal conductivity of the plane wall material in W/(m-K). k is assumed to be close to a constant in any direction within the plane wall. For definition of characteristic length of a solid volume, see Chapter 2. h is the constant convection or radiation heat transfer coefficient in W/(m^2-K) where this constant value can be used in a small interval of time in our transient calculations. Second non-dimensional number is the Fourier number given in Equation 19-2. The Fourier number is the non-dimensional time which appears in all transient heat transfer calculations.

$$Biot\ \#(BI) = \frac{(L/k)}{(1/h)} \qquad 19\text{-}1$$

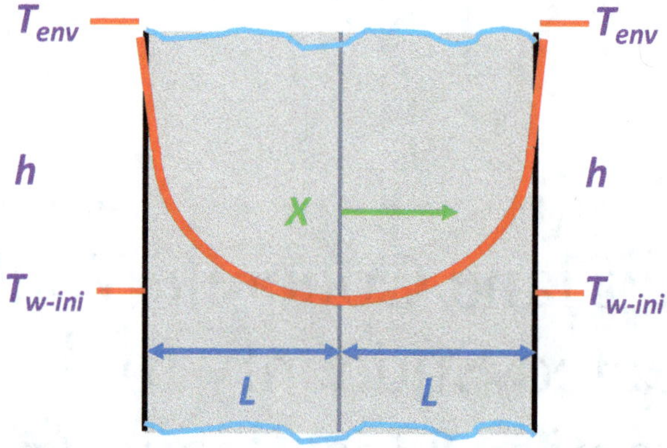

Figure 19-1 Large plane concrete wall with a **2*L*** width is subjected suddenly to convection heat transfer ***h*** on both sides and $T_{w-ini} < T_{env}$. Temperature distribution within the plane wall is a function of ***X*** and θ.

$$Fourier\ \#(FO) = \frac{\alpha\theta}{L^2} \tag{19-2}$$

$\alpha = \dfrac{k}{\rho C_p}$ is the thermal diffusivity of the plane wall material in m²/s. ρ is the density of the plane wall material in kg/m³. C_p is the specific heat of the plane wall material at constant pressure in J/(kg-K). θ is the time in seconds.

If ***BI*** < **0.1**, temperature-gradients within the plane wall can be neglected and the energy balance for the plane wall gives us the following differential equation and its initial condition in Equations 19-3.

$$\frac{dT}{d(FO)} = -BI\left(T - T_{env}\right) \tag{19-3a}$$

$$T(0) = T_{w-ini} - T_{env} \tag{19-3b}$$

Separating variables and integrating Equation 19-3a and applying the initial condition Equation 19-3b give the lumped capacitance method solution shown in Equation 19-4 for determining temperature in a large plane wall which varies only with time and temperature gradients within the plane wall are negligible.

$$\frac{T(\theta) - T_{env}}{T_{w-ini} - T_{env}} = exp(-BI \times FO) \tag{19-4}$$

The lumped capacitance method which is given in Equation 19-4 does not apply when $\boldsymbol{BI > 0.1}$. In order to determine the temperature distribution within the large plane wall, we have to introduce the X coordinate into the conservation of energy equation. A large plane wall is suddenly subjected to convection heat transfer at its $X = \pm L$ surfaces. The partial differential equation for the transient temperature distribution, $T(X,\theta)$, with constant wall properties and with no internal heat generation with its initial and boundary conditions are shown in Equations 19-5.

$$\frac{1}{\alpha}\frac{\partial T}{\partial \theta} = \frac{\partial^2 T}{\partial X^2} \qquad \text{19-5a}$$

$$T(X,0) = T_{w-ini} \qquad \text{19-5b}$$

$$\left(\frac{\partial T}{\partial X}\right)_{X=0} = 0 \qquad \text{19-5c}$$

$$-k\left(\frac{\partial T}{\partial X}\right)_{X=L} = h\left[T(\pm L,\theta) - T_{env}\right] \qquad \text{19-5d}$$

In Equations 19-5, $T(X,\theta)$, is the transient temperature in X location in meters and time, θ, in seconds. Exact solutions to Equations 19-5 are given in Equations 19-6, for details see Ref. [15].

$$T(X,\theta) = \frac{T(X,\theta) - T_{env}}{T_{w-ini} - T_{env}} = \sum_{i=1}^{\infty} A_i exp^{\left(-\lambda_i^2 FO\right)} cos\left(\lambda_i \frac{X}{L}\right) \qquad \text{19-6a}$$

$$BI = \lambda_i tan(\lambda_i) \qquad \text{19-6b}$$

$$A_i = \frac{4sin(\lambda_i)}{2\lambda_i + sin(2\lambda_i)} \qquad \text{19-6c}$$

There is also an approximate solution to Equations 19-5, when $\boldsymbol{FO > 0.2}$. The dominant term in the series summation in Equation 19-6a is the $i = 1$ term. When $FO > 0.2$, only the first term in Equation 19-6 can be used as shown in Equations 19-7.

$$T(X,\theta) = \frac{T(X,\theta) - T_{env}}{T_{w-ini} - T_{env}} = A_1 exp^{\left(-\lambda_1^2 FO\right)} cos\left(\lambda_1 \frac{X}{L}\right) \qquad \text{19-7a}$$

$$BI = \lambda_1 tan(\lambda_1) \qquad \text{19-7b}$$

$$A_1 = \frac{4sin(\lambda_1)}{2\lambda_1 + sin(2\lambda_1)} \qquad \text{19-7c}$$

Using following parameters, large plane concrete wall's centerline temperature at $X = 0$ can be calculated using the lumped capacitance method, namely Equation 19-4, the exact solution by using first five terms of Equations 19-6, or the approximation of the exact solution, namely Equations 19-7.

$$T_{w-ini} = 25C; \qquad T_{env} = 100C$$

$$L = 0.03\,m; \qquad k = 1\,W/(m-K)$$

$$\alpha = 5.56\times10^{-7}\ m^2/s; \qquad h = 10\ W/(m^2-K)$$

Figure 19-2 shows the large plane concrete wall's centerline temperature versus time using the exact solution using first five terms from Equation 19-6a, using one-term approximation of the exact solution using Equation 19-7a and the lumped capacitance method using Equation 19-4 for $BI = 0.3$. Centerline temperature is underestimated by using the one-term approximation up to 330 seconds or $FO = 0.2$. The lumped capacitance method overestimates the centerline temperature from the exact results which verifies that temperature gradients within large plane concrete wall are significant. Figure 19-3 compares exact and approximate centerline temperatures at two different Biot numbers, namely $BI = 1.2$ *and* 30. This graph shows the effects of convection heat transfer coefficient increasing from 40 $W/(m^2-K)$ to 1,000 $W/(m^2-K)$ to plane wall's centerline temperature. For $\theta > 324\ s$ or $FO > 0.2$, one-term approximation is a good one in place of the five-term exact solution.

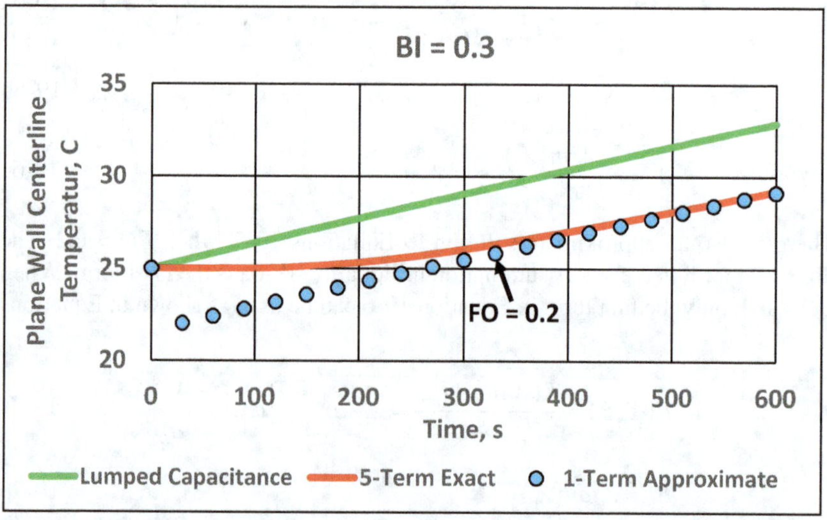

Figure 19-2 Large plane concrete wall's centerline temperature versus time using three methods: The lumped capacitance method, the exact solution using first five terms, and the one term approximation to the exact solution

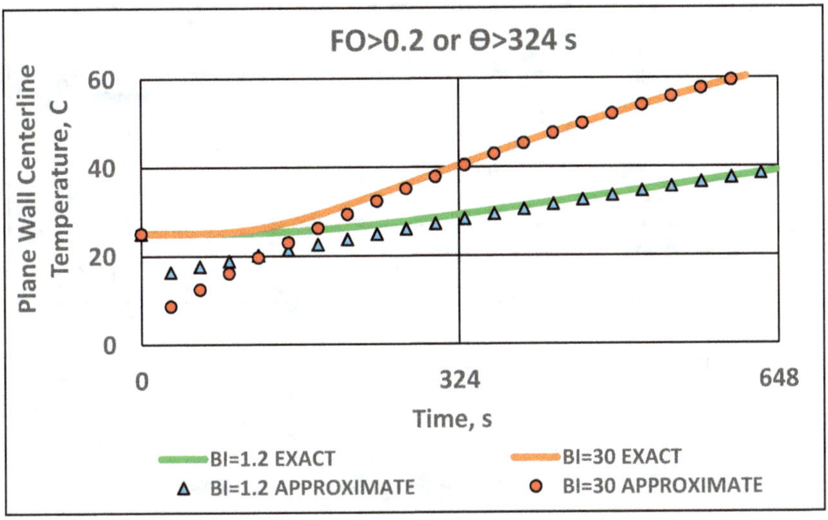

Figure 19-3 Exact and approximate centerline temperatures versus time at two different Biot numbers 1.2 and 30

Three dimensionless numbers, *BI*, *FO*, and *BIFO* cover all different transient temperature distribution solutions for a large plane wall subjected suddenly to convection heat transfer at both of its surfaces. *BIFO* for the lumped capacitance method is shown in Equation 19-8.

$$BIFO = \frac{h\theta}{\rho C_p L} = \frac{\theta}{\tau_{LC}} \qquad 19\text{-}8$$

where τ_{LC} is characterized as the time constant for thermal energy distribution in the plane wall in seconds for nearly uniform temperatures in the solid wall. τ_{LC} is the ratio of plane wall material's ability to store thermal energy to convection heat transfer coefficient at plane wall's surfaces. High τ_{LC} means that the plane wall material can store lots of thermal energy and the convection heat transfer coefficient is low and therefore the temperature inside the wall changes slowly.

Figure 19-4 shows plane wall's non-dimensional centerline temperature, $T^* = \frac{T_{w-cl} - T_{env}}{T_{w-ini} - T_{env}}$, versus BI number at five different FO values. As BI increases, thermal energy coming into the plane wall from convection heat transfer increases and therefore, plane wall's non-dimensional centerline temperature, T^*, approaches zero faster.

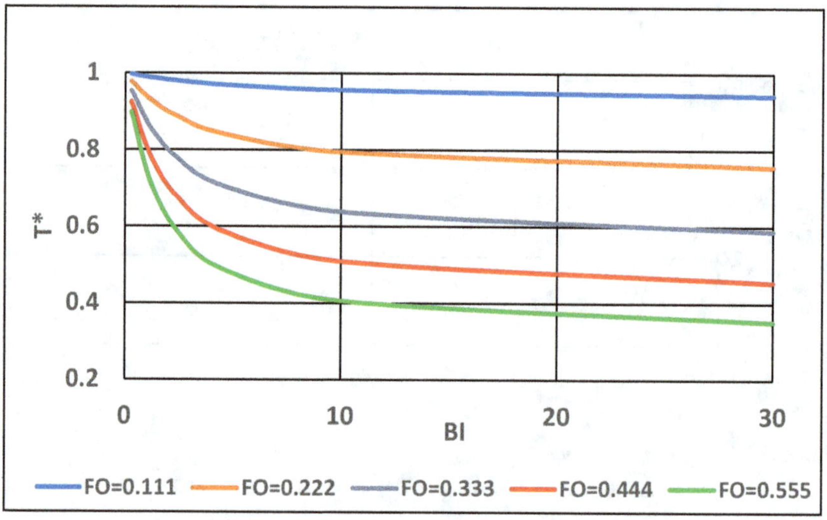

Figure 19-4 Concrete plane wall's non-dimensional centerline temperature versus *BI* number at five different *FO* values

Sensitivities of T^* to BI, $\partial(T^*)/\partial(BI)$ versus BI for different times, *FO*, are shown in Figure 19-5. As time increases, absolute value of $\partial(T^*)/\partial(BI)$ gets larger. However, sensitivity of T^* to *BI*, approaches zero fast for $BI > 20$, namely all centerline temperature versus-*BI* profiles have almost the same slope for $BI > 20$.

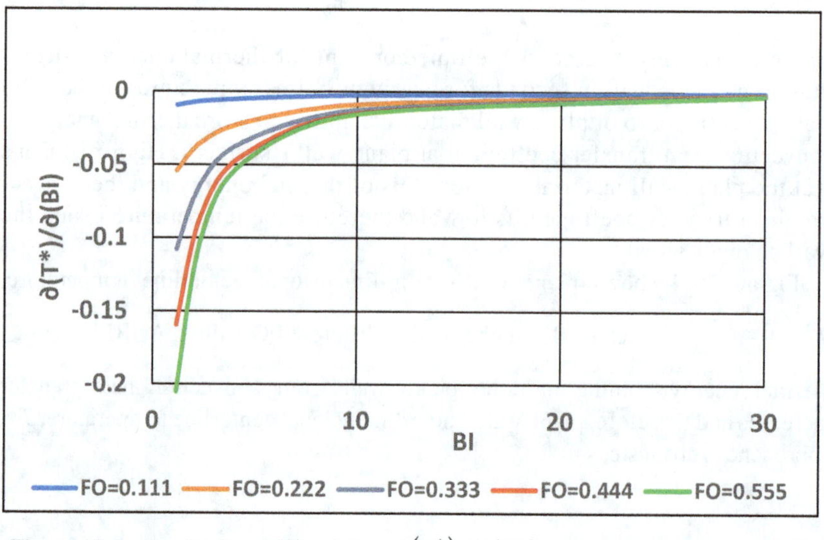

Figure 19-5 Sensitivity of T^* to BI or $\partial(T^*)/\partial(BI)$ versus *BI* for different *FO's*

Figure 19-6 shows the concrete plane wall's non-dimensional centerline temperature versus FO number at six different BI values. As BI increases, thermal energy coming into the plane wall from convection heat transfer increases and therefore, plane wall's non-dimensional centerline temperature, T^*, approaches zero faster. Sensitivities of T^* to FO, $\partial(T^*)/\partial(FO)$ versus FO for six different BI values are shown in Figure 19-7. Absolute values of $\partial(T^*)/\partial(FO)$ maximize at shorter times as *BI* increases. This shows that as the thermal energy in convection heat transfer at plane wall's surfaces increases, the change in centerline temperature maximizes at shorter times.

A concrete plane wall's non-dimensional centerline-temperature, T^*, versus *BIFO* using the exact solution with five terms and also using the lumped capacitance method are shown in Figure 19-8. For $BIFO = 10$, T^*, using the five-term exact method, approaches 0.8. When $BIFO = 5$, however, T^*, using the lumped capacitance method, approaches the environment temperature. This is because the lumped capacitance method does not consider conduction heat transfer within the plane wall which distributes incoming thermal energy within the wall before it arrives at its centerline. Figure 19-8 also shows that for $BIFO = \frac{\theta}{\tau} < 0.3$, the five-term exact method and the lumped capacitance method give similar results.

For example, at a low convection heat transfer coefficient of 10 W/(m^2-K), or BI = 0.3, $\tau = 5{,}400\ s$. Then, at $BIFO = 5$, $T^* = 0.011$ or $T_{w-cl} = 99.2\ C$. It takes $5{,}400 \times 5 = 27{,}000\ s$ for the plane wall's centerline temperature to reach almost the environment's temperature.

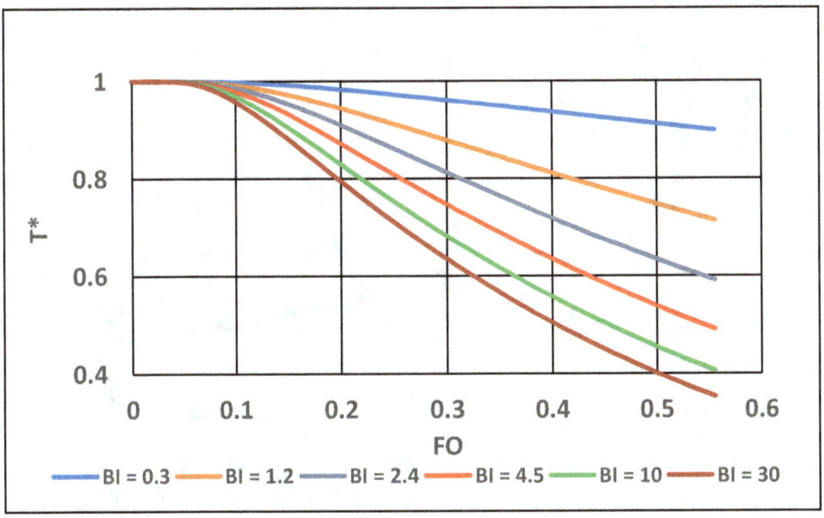

Figure 19-6 Concrete plane wall's non-dimensional centerline temperature versus *FO* number at six different *BI* values

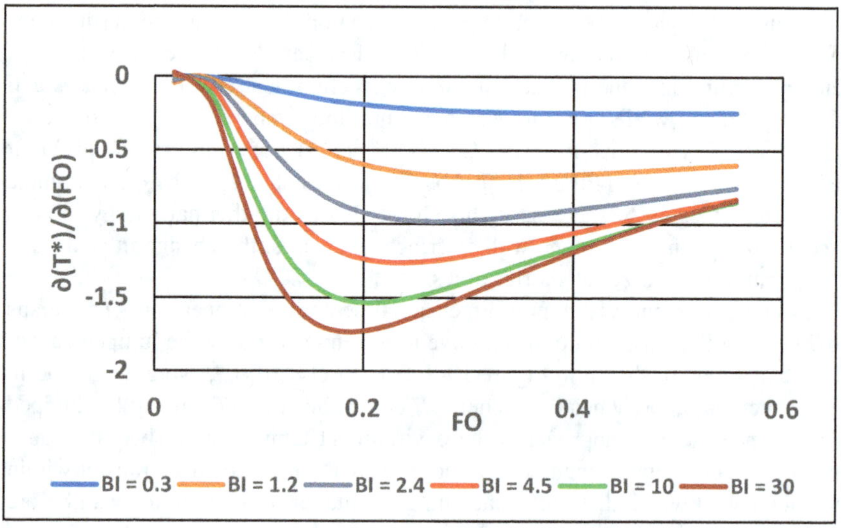

Figure 19-7 Sensitivity of T^* to FO or $\partial(T^*)/\partial(FO)$ versus FO for six different BI values

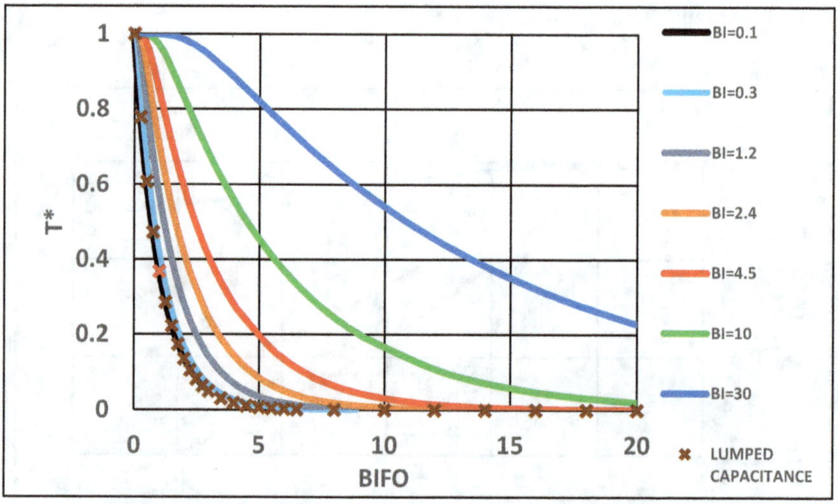

Figure 19-8 A concrete plane wall's non-dimensional centerline temperature, T^*, versus $BIFO$ using the five-term exact solution and the lumped capacitance method

At a high convection heat transfer coefficient of 1,000 W/(m²-K), or BI = 30, $\tau = 54\ s$. Then, at *BIFO* = 5, *T** = 0.82 or T_{w-cl} = 38.5 *C*. It takes 54 × 5 = 270 *s* for the plane wall's centerline temperature to reach 38.5 C temperature. the plane wall's centerline temperature to reaches 99.2 C temperature at *BIFO* = 64.8 or at 54 × 64.8 = 3,500 *s*.

Figure 19-9 shows a concrete plane wall's non-dimensional centerline temperature, *T**, versus BIFO using the five-term exact solution and the lumped capacitance method are for *BIFO* ≤ 5. *BI* = 0.1 and *BI* = 0.3 curves follow the lumped capacitance method curve, Equation 19-4, very closely for 0 ≤ *T** ≤ 1.

Figure 19-10 shows the sensitivity of *T** to BIFO or ∂(*T**)/ ∂(*BIFO*) versus *BIFO* for different *BI* numbers or convection heat transfer coefficients.

Absolute values of sensitivities of centerline temperature to BIFO decrease fast at low convection heat transfer coefficients, i.e. $h = 10 \dfrac{W}{m^2 - K}$ *or BI* = 0.3. Sensitivities of low *BI* curves follow sensitivity of the lumped capacitance method curve very closely.

Figure 19-11 shows the sensitivity of centerline temperature, *LN*(*T**), to *BIFO* or ∂[*LN*(*T**)]/∂(*BIFO*) versus *BIFO* for different *BI* numbers at constant slope regions. Slopes of these curves are constants using the first term approximation in Equation 19-7, namely $-\dfrac{\lambda_1^2}{BI} = -\dfrac{\lambda_1}{tan(\lambda_1)}$, are shown in Table 19.1. This sensitivity is -1 for the lumped capacitance method and its absolute value decreases with increasing *BI*.

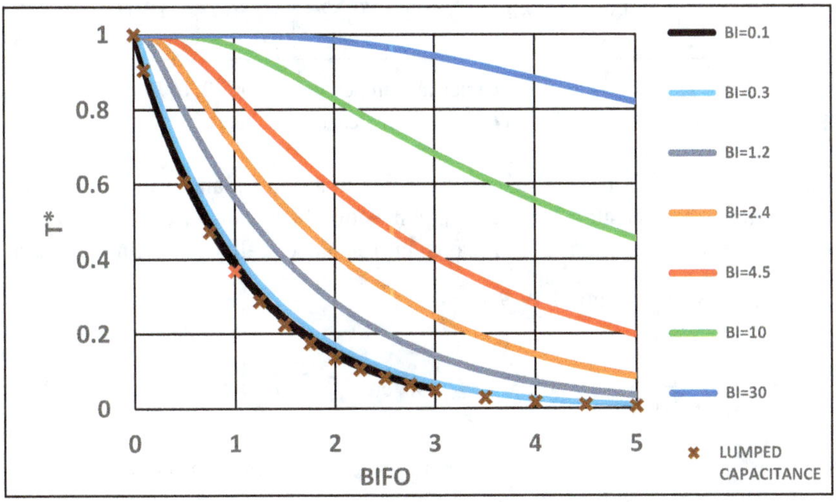

Figure 19-9 A concrete plane wall's non-dimensional centerline temperature, *T**, versus *BIFO* using the five-term exact solution and the lumped capacitance method are shown for *BIFO* ≤ 5.

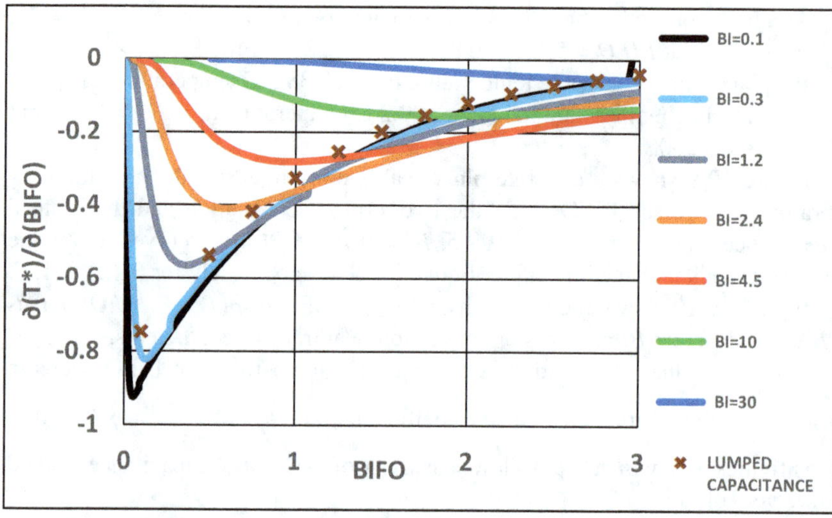

Figure 19-10 Sensitivity of centerline temperature, T^*, to *BIFO* or $\partial(T^*)]/\partial(BIFO)$ versus *BIFO* for different *BI* numbers

h, W/(m²-K)⁾	3.33	10	40	80	150	333.3	1,000
BI	0.1	0.3	1.2	2.4	4.5	10	30
$\frac{\lambda_1^2}{BI}$	-0.968	-0.908	-0.692	-0.526	-0.369	-0.204	-0.077

Table 19-1 Sensitivity of centerline temperature, $LN(T^*)$, to *BIFO* or $\partial[LN(T^*)]/\partial(BIFO)$ for increasing numbers

Using Equation 19-7 for the first-term approximation of plane wall temperature distribution which is subject to sudden convection heat transfer at both surfaces, we can define a new thermal time constant for plane wall's centerline temperature, τ_{CL}, shown in Equation 19-9. $\frac{\theta}{\tau_{CL}}$

$$T(X,\theta) = A_1 exp^{\left(-\frac{\theta}{\tau_{CL}}\right)} \quad \text{where} \quad \tau_{CL} = \frac{L^2}{\alpha\lambda_1^2} \qquad 19\text{-}9$$

Next, let us investigate transient temperature distributions in a large plane concrete wall which is subjected suddenly to convection heat transfer on both surfaces using the explicit finite difference method with parameters given above. The only drawback in the explicit finite difference method is the stability in calculations. Maximum time increment for calculations has to be determined in order

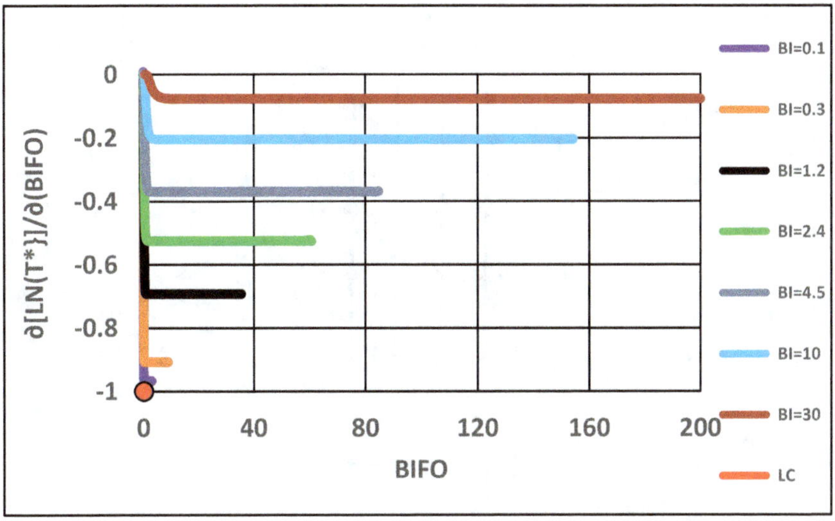

Figure 19-11 Sensitivity of centerline temperature, $LN(T^*)$, to $BIFO$ or $\partial[LN(T^*)]/\partial(BIFO)$ versus BIFO

to achieve stable results. Let us redraw Figure 19-1 as shown in Figure 19-12 to determine explicit finite difference equations for a large plane concrete wall which is subjected suddenly to convection heat transfer on both surfaces. Since convection heat transfer conditions at $X = \pm L$ are the same and temperatures with in the wall are symmetrical, only the section of the wall from $X = 0$ to $X = L$ will be considered and the symmetry line at $X = 0$ will be insulated.

For the explicit finite difference method, as a result of energy balance at a node, BI and FO have to be redefined as the length between two nodes, namely with ΔX, as given in Equations 19-10.

$$BI = \frac{h\Delta X}{k} \tag{19-10a}$$

$$FO = \frac{\alpha\Delta\theta}{(\Delta X)^2} \tag{19-10b}$$

The following Equation 19-11 results from energy balance for node 1 in Figure 19-12.

$$\rho C_p\left(\frac{A\Delta X}{2}\right)\frac{\left(T_1^{j+1} - T_1^j\right)}{\Delta\theta} = \left(\frac{kA}{\Delta X}\right)\left(T_2^j - T_1^j\right) \tag{19-11}$$

where A is the surface area between T_1 and T_2 in m^2 and $\frac{A\Delta X}{2}$ is the volume of T_1.

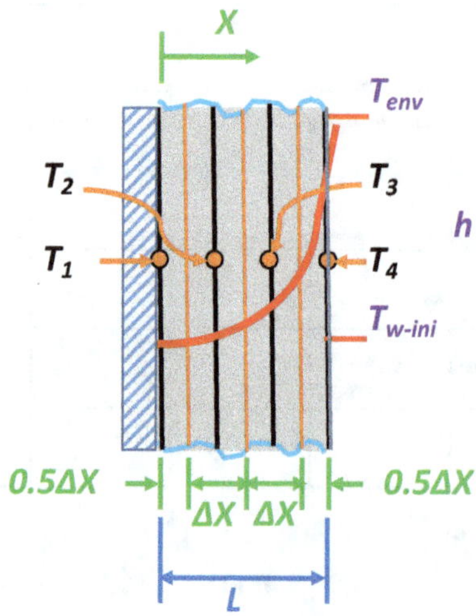

Figure 19-12 For the explicit finite difference method, only four temperature nodes, T_1, T_2, T_3, and T_4, within half of the plane wall are shown.

Equation 19-11 can be rewritten in the following more convenient form as shown in Equation 19-12

$$T_1^{j+1} = (1-2FO)T_1^j + 2FOT_2^j \qquad 19\text{-}12$$

The following Equation 19-13 which results from energy balance for node 2 is shown Figure 19-11.

$$\rho C_p\left(\frac{A\Delta X}{2}\right)\frac{\left(T_2^{j+1}-T_2^j\right)}{\Delta\theta} = \left(\frac{kA}{\Delta X}\right)\left(T_1^j - T_2^j\right) + \left(\frac{kA}{\Delta X}\right)\left(T_3^j - T_2^j\right) \qquad 19\text{-}13$$

Equation 19-13 can be rewritten in the following more convenient form as shown in Equation 19-14.

$$T_2^{j+1} = (1-2FO)T_2^j + FO\left(T_1^j + T_3^j\right) \qquad 19\text{-}14$$

The following Equation 19-15 results from energy balance for node 3 which is similar to node 2 as shown in Figure 19-12.

$$\rho C_p\left(\frac{A\Delta X}{2}\right)\frac{\left(T_3^{j+1}-T_3^j\right)}{\Delta\theta} = \left(\frac{kA}{\Delta X}\right)\left(T_2^j - T_3^j\right) + \left(\frac{kA}{\Delta X}\right)\left(T_4^j - T_3^j\right) \qquad 19\text{-}15$$

Equation 19-15 can be rewritten in the following more convenient form as shown in Equation 19-16.

$$T_3^{j+1} = (1-2FO)T_3^j + FO\left(T_2^j + T_4^j\right) \qquad 19\text{-}16$$

The following Equation 19-17 results from energy balance for node 4 in Figure 19-12.

$$\rho C_p\left(\frac{A\Delta X}{2}\right)\frac{\left(T_4^{j+1}-T_4^j\right)}{\Delta\theta}=hA\left(T_{env}-T_4^j\right)+\left(\frac{kA}{\Delta X}\right)\left(T_3^j-T_4^j\right) \qquad 19\text{-}17$$

Equation 19-17 can be rewritten in the following more convenient form as shown in Equation 19-18.

$$T_4^{j+1}=\left(1-2BIFO-2FO\right)T_4^j+2FOT_3^j+2BIFOT_{env} \qquad 19\text{-}18$$

In Equations 19-11, 19-13, 19-15 and 19-17, temperatures at time $j + 1$ are calculated from known temperatures at previous time j. However, in order to have stable results, time interval, $\Delta\theta$, used in calculations has to be such that coefficients of T_1^j, T_2^j, T_3^j and T_4^j have to be greater than zero as shown in Equations 19-18 and 19-19.

For Equations 19-12, 19-14 and 19-16, the stability criterion using following parameters is shown in Equation 19-19.

$T_{env}=100C$ $\quad L=0.03\,m$ $\quad \Delta X=0.01\,m$;

$k=1\ W/\left(m-K\right)$ $\quad \alpha=5.56\times10^{-7}\ m^2/s$;

$h=10\ W/\left(m^2-K\right)$

$$1-2FO\geq 0 \ \text{ or } \ FO\leq 0.5 \ \text{ or } \ \Delta\theta\leq\frac{0.5\left(\Delta X\right)^2}{\alpha}=90\,s \qquad 19\text{-}19$$

For Equation 19-18, the stability criterion is shown in Equation 19-20 and for different h values they are tabulated in Table 19-2.

$$1-2FO\left(1+BI\right)\geq 0 \qquad FO\leq\frac{0.5}{1+BI} \qquad 19\text{-}20$$

h, W/(m^2-K)	BI	$FO \leq 0.5/(1+BI)$	$\Delta\theta \leq$, s
10	0.1	0.454	81.8
40	0.4	0.357	64.2
80	0.8	0.277	50.0
150	1.5	0.200	36.0
333.3	3.33	0.115	20.8
1000	10.0	0.045	8.18

Table 19-2 Stability criteria are shown for different ***h*** values in Equation 19-20

Results in Table 19-2 shows that as thermal energy in convection heat transfer increases, time increment, $\Delta\theta$, decreases in order to have stable results from explicit finite difference method. For example, a time increment of 80 seconds was used for $BI = 0.1$ *or* $h = 10\ W/(m^2 - K)$ explicit finite difference calculations and a time increment of 8 seconds was used for $BI = 10$ *or* $h = 1{,}000\ W/(m^2 - K)$ explicit finite difference calculations.

Figure 19-13 shows comparisons of large plane concrete wall centerline temperatures versus time between five-term exact solutions from Equations 19-6 and explicit finite difference solutions discussed above for different BI numbers. Temperature-time distributions follow each other very closely when both calculation methods are used. When stability criteria are followed accurately, explicit finite difference method is a very good alternative calculation method. Especially in complicated geometries where an exact solution is impossible, explicit finite difference method can be used.

If time increment requirements in an explicit finite difference system of equations is too small and solutions require large amount of computer times, then implicit finite difference can be utilized. In the implicit finite difference method, the future temperature, $(j + 1)$, at a node depends on the present temperature, (j), only at the same node in the transient term, namely in $\frac{\left(T_n^{j+1} - T_n^j\right)}{\Delta\theta}$, and depends on present temperatures of all other surrounding nodes. Implicit finite difference

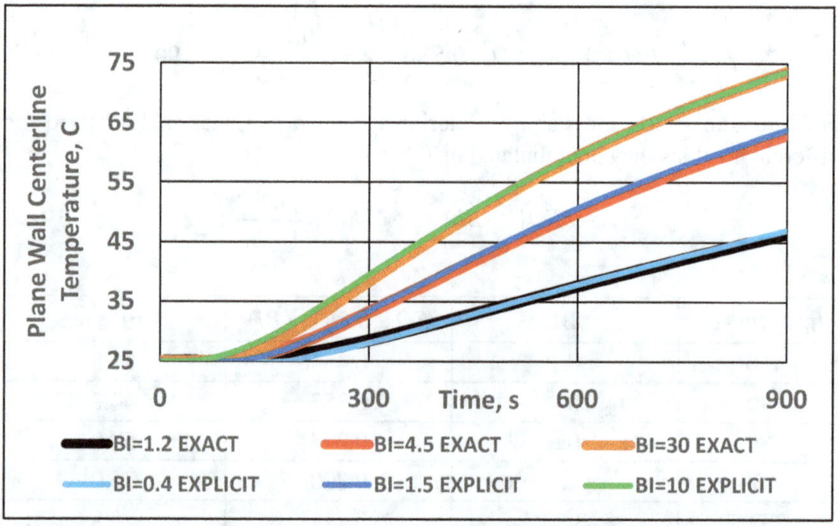

Figure 19-13 Large plane concrete wall centerline temperature versus time obtained from five-term exact solutions using Equations 19-6 and explicit finite difference solutions for different BI numbers

method does not require stability criteria in time increments. However, there might be small space increment requirements between nodes and small-time step requirements in order to increase accuracy of results.

Implicit finite difference method's energy balances are similar to explicit finite difference method's energy balances. All temperatures with j time exponent become $j + 1$ exponent except in the transient term $\frac{\left(T_n^{j+1} - T_n^j\right)}{\Delta\theta}$. Using the four-node system in Figure 19-12, we get the following Equations 19-21, 19-22, 19-23, and 19-24.

For node 1:

$$(1+2FO)T_1^{j+1} - 2FOT_2^{j+1} = T_1^j \qquad 19\text{-}21$$

For node 2:

$$(1+2FO)T_2^{j+1} - FOT_1^{j+1} - FOT_2^{j+1} = T_2^j \qquad 19\text{-}22$$

For node 3:

$$(1+2FO)T_3^{j+1} - FOT_2^{j+1} - FOT_4^{j+1} = T_3^j \qquad 19\text{-}23$$

For node 4:

$$(1+2FO+2BIFO)T_4^{j+1} - 2FOT_3^{j+1} = T_4^j + 2BIFOT_{env} \qquad 19\text{-}24$$

In Equations 19-21, 19-22, 19-23 and 19-24, temperatures at time $j + 1$ can be calculated by matrix inversion method as shown below.

$j + 1$ temperatures have constant coefficients in a 4x4 matrix $[A]$.

$$[A] = \begin{vmatrix} (1+2FO) & -2FO & 0 & 0 \\ -FO & (1+2FO) & -FO & 0 \\ 0 & -FO & (1+2FO) & -FO \\ 0 & 0 & -2FO & (1+2FO+2BIFO) \end{vmatrix}$$

$$[\boldsymbol{A}][\boldsymbol{T}]=[\boldsymbol{B}]$$

and

where $[\boldsymbol{T}] = \begin{bmatrix} T_1^{j+1} \\ T_2^{j+1} \\ T_3^{j+1} \\ T_4^{j+1} \end{bmatrix}$ and $[\boldsymbol{B}] = \begin{bmatrix} T_1^j \\ T_2^j \\ T_3^j \\ T_4^j + 2BIFOT_{env} \end{bmatrix}$

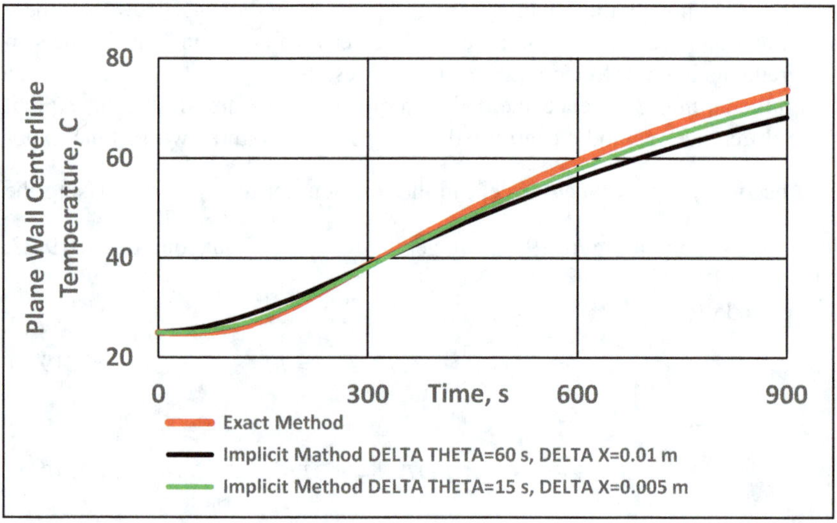

Figure 19-14 Comparisons of large plane concrete wall centerline temperatures versus time between five-term exact solution from Equations 19-6 and implicit finite difference solutions for two cases. In one case, $\Delta\theta = 60\ s$ and $\Delta X = 0.01\ m$ and in the other case $\Delta\theta = 15\ s$ and $\Delta X = 0.005\ m$

After inversion of matrix **[*A*]**, **[*T*]** = **[*A*]**$^{-1}$ **[*B*]**. Beware that, matrix **[*B*]** changes at each time step of calculations.

Figure 19-14 shows comparisons of large plane concrete wall centerline temperatures versus time between five-term exact solution from Equations 19-6 and implicit finite difference matrix inversion solution to Equations 19-21 through 19-24. With smaller space increments between nodes and with smaller time steps, centerline temperature versus time calculated by the implicit finite difference method gets very close to the five-term exact solution. We did not encounter any stability issues in calculations using the implicit finite difference method.

20

Seasonally Averaged Temperature Fluctuation Penetrations into The Ground

It can be very complicated to determine soil temperatures with changing depth and time. Many factors such as soil composition and porosity, running, seeping or stationary water, ice, vegetation, trees, rain, snow, day and night temperature variations, global warming and human intervention, etc., affect soil temperature penetration. In this chapter, we will investigate seasonally averaged varying temperature effects on a specific type of soil. We will assume an average and constant surface temperature over a specific type of soil over 90 days. The average and constant surface temperature of soil in Winter months of December, January and February will be -4.44 C. The average and constant surface temperature of soil in Spring months of March, April and May will be +3.33 C. The average and constant surface temperature of soil in the Summer months of June, July and August will be +15.0 C. The average and constant surface temperature of soil in Fall months of September, October and November will be +5.55 C.

Following partial differential equation and initial and boundary conditions apply to a semi-infinite solid soil, which is suddenly exposed to constant surface temperatures given above for each season, are shown in Equations 20-1.

$$\frac{\partial^2 T}{\partial X^2} = \frac{1}{\alpha}\frac{\partial T}{\partial \theta} \qquad \text{20-1a}$$

$$T(X,0) = T_i \qquad \text{20-1b}$$

$$T(X \to \infty, \theta) = T_i \quad \text{20-1c}$$

$$T(0, \theta) = T_{soil\ surface} \quad \text{20-1d}$$

$T(X,\theta)$ is the temperature in the semi-infinite solid soil at the location in meters from its surface and at a time in seconds. The temperature distribution in a semi-infinite solid soil that satisfies Equations 18-1 is shown in Equation 20-2. The heat transfer at the surface of the semi-infinite solid, $X = 0$, is given in Equation 20-3.

$$\frac{T(X,\theta) - T_{int}}{T_{i-1} - T_{int}} = erf\left(\frac{X}{2\sqrt{\alpha\theta}}\right) \quad \text{20-2}$$

$$q(0,\theta) = \frac{k(T_{i-1} - T_{int})}{\sqrt{\pi\alpha\theta}} \quad \text{20-3}$$

where k is semi-infinite solid's thermal conductivity in W/(m-K) and α is semi-infinite solid's thermal diffusivity in m²/s. $\left(\frac{X}{2\sqrt{\alpha_s\theta}}\right)$ is the dimensionless similarity variable that converts the partial differential into an ordinary one. See Chapter 15 for details of the Gaussian error function $erf\left(\frac{X}{2\sqrt{\alpha_s\theta}}\right)$.

Following parameters are used to determine soil temperatures with changing depth and time under each season's average temperatures. Average temperatures used as soil's constant surface temperature for each season are given below.

$T_{soil\ surface-winter} = -4.44\ C$; $\quad T_{soil\ surface-spring} = 3.33\ C$

$T_{soil\ surface-summer} = 15.0\ C$; $\quad T_{soil\ surface-fall} = 5.55\ C$

$\alpha_s = 1.4 \times 10^{-7}\ m^2/s$

Depths at which soil temperatures are calculated:

$X = 0, 0.1, 0.2, 0.3, 0.4, 0.5, 0.6, 0.7, 0.8, 0.9,\ and\ 1.0\ m$

Initial temperature for each soil depth location in a new season is the 90th day temperature from the last season at that same soil depth location. For example:

$$T_{i-spring\ at\ X=0.5} = T_{winter\ at\ X=0.5\ and\ \theta=90\ days}$$

Winter season soil temperature profiles for 90 days at different depths up to 1.0 meter are shown in Figure 20-1. Sensitivities of soil temperatures to time, $\frac{\partial T}{\partial \theta}$, decrease as depth increases and as time increases. Depths at $X = 0.8\ m$ and below, soil temperatures do not reach the freezing temperature point of zero degree during 90 days of Winter season.

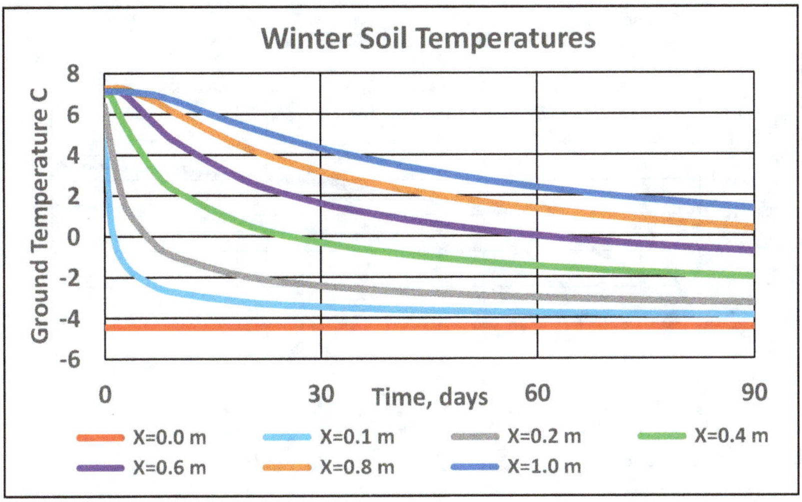

Figure 20-1 Winter season soil temperature profiles for 90 days at different depths

Spring season soil temperature profiles for 90 days for different depths up to 1.0 meter are shown in Figure 20-2. Frozen depth temperatures such as at $X = 0.4\ \ m\ and\ 0.6\ m$, are very sensitive, namely $\frac{\partial T}{\partial \theta}$, to constant and positive Spring surface temperature up to one week into spring. After one week, all depths get out of the frozen zone of zero degree. Sensitivities at all depths, or $\frac{\partial T}{\partial \theta}$, up to $X = 1.0\ \ m$ diminish as the end of Spring is approached.

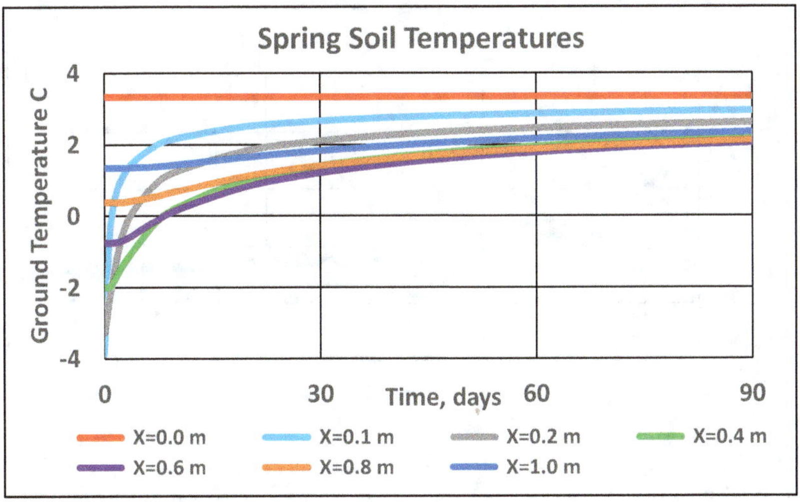

Figure 20-2 Spring season soil temperature profiles for 90 days at different depths

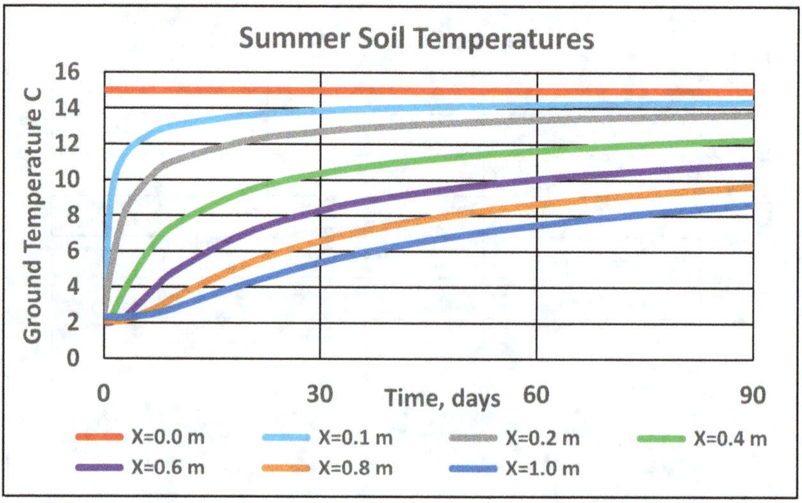

Figure 20-3 Spring season soil temperature profiles for 90 days at different depths

Summer season soil temperature profiles for 90 days for different depths up to 1.0 meter are shown in Figure 20-3. Sensitivities, namely $\frac{\partial T}{\partial \theta}$, to constant surface temperature of 15 C decrease as depth increases. As the end of Summer is approached, sensitivities, or $\frac{\partial T}{\partial \theta}$, at all depths up to 1.0-meter approach zero.

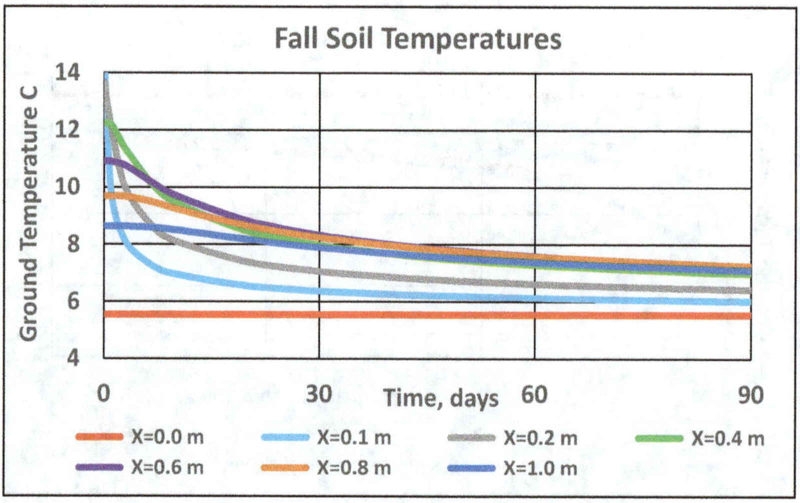

Figure 20-4 Fall season soil temperature profiles for 90 days at different depths

Fall season soil temperature profiles for 90 days for different depths up to 1.0 meter are shown in Figure 20-4. Soil depth temperatures such as at $X = 0.4\ m$ *and* 0.6 *m*, are sensitive, namely $\frac{\partial T}{\partial \theta}$, to constant and positive Fall surface temperature of 5.55 C up to one month into Fall. Sensitivities at all depths, or $\frac{\partial T}{\partial \theta}$, up to $X = 1.0\ m$ diminish as the end of Fall is approached.

Figure 20-5 shows the minimum, the maximum and the average temperatures at $X = 1.0\ m$ depth during four seasons. During heating of ground soil which is close to the surface, at $X = 1.0\ m$, during Spring and Summer seasons, average temperatures follow Spring and Summer seasons' maximum temperatures very closely.

Figure 20-6 shows the minimum, the maximum and the average temperatures at $X = 0.5\ m$ depth during four seasons. At this depth, minimum, average and maximum temperatures are evenly spaced, except at the end of Summer and at the beginning of the Fall seasons.

At the end of Summer and at the beginning of the Fall seasons, minimum temperatures at $X = 0.5\ m$ warm up and get closer to $X = 0.5\ m$ average temperature.

Figure 20-7 shows the minimum, the maximum and the average temperatures at depth during four seasons. At this depth, minimum, average and maximum temperatures are very close to each other because effects of soil surface temperatures seasonally averaged penetrations into ground soil are diminishing at this depth.

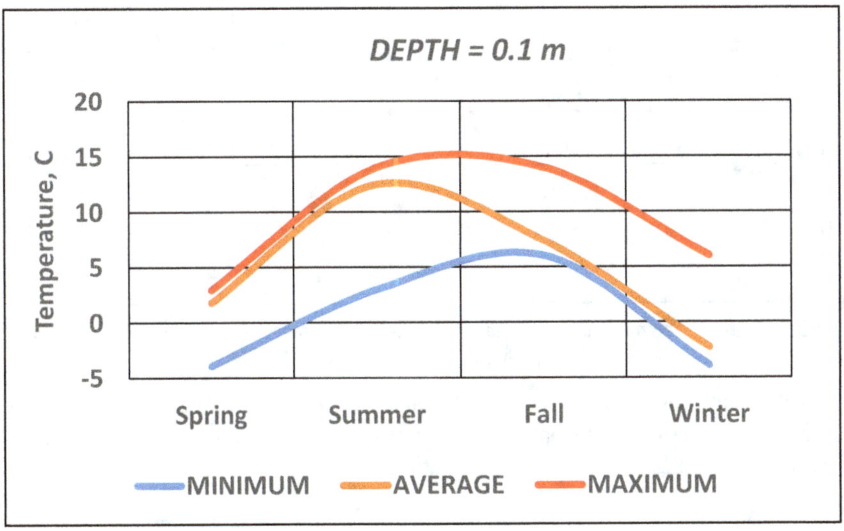

Figure 20-5 The minimum, the maximum and the average temperatures $X = 0.1\ m$ at depth during four seasons

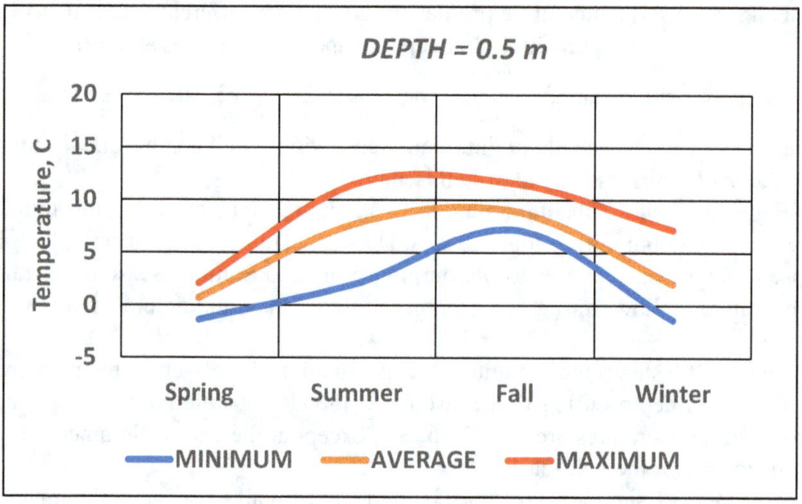

Figure 20-6 The minimum, the maximum and the average temperatures at $X = 0.5\ m$ depth during four seasons

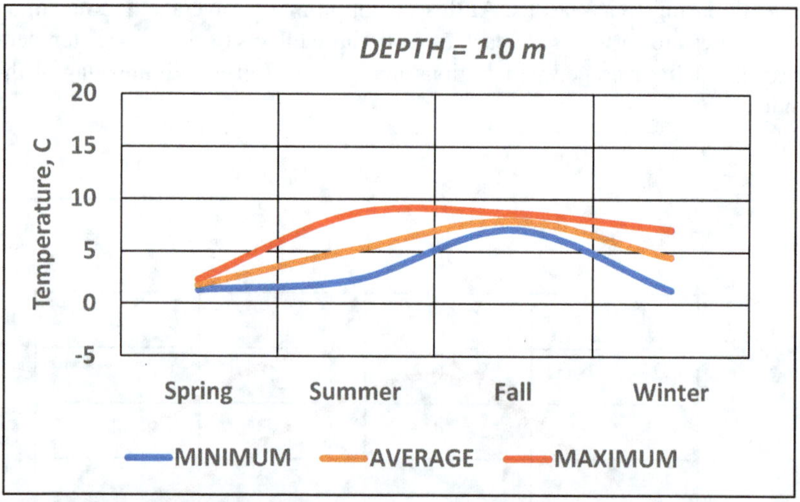

Figure 20-7 The minimum, the maximum and the average temperatures at $X = 1.0\ m$ depth during four seasons

Total heat transfer per unit area, *Q/A* in W/m^2, in and out of soil's surface can be calculated for each season, or for 90 days or for 7,776,000 seconds, by integrating Equation 20-3 as shown in Equation 20-4.

$$Q / A = \int_{0}^{7,776,000} \frac{k\left(T_{soil\ surface} - T_i\right)}{\sqrt{\pi\alpha\theta}} d\theta = 2k\left(T_{soil\ surface} - T_i\right)\sqrt{\frac{7,776,000}{\pi\alpha}} \quad 20\text{-}4$$

where k is the constant thermal conductivity of soil at 0.5 W/(m-K). Winter season Q/A is out of soil's surface at $-4.40\times10^{+7}$ W/m^2. Spring season Q/A is into soil's surface at $+3.03\times10^{+7}$ W/m^2. Summer season Q/A is into soil's surface at $+5.06\times10^{+7}$ W/m^2. Fall season Q/A is out of soil's surface at-$3.70\times10^{+7}$ W/m^2. After a one-year seasonal cycle, net total heat transfer per unit area in and out of soil's surface is zero.

21

Cooling of a Desert's Surface on Earth at Night from Infrared Radiation Heat Transfer

In this chapter, we will investigate transient infrared radiation heat transfer at night from a desert's surface at $X = 0$ which is suddenly exposed to outgoing and incoming constant infrared radiation heat flux. Solution to one-dimensional transient conduction heat transfer in a semi-infinite solid which is suddenly exposed to a constant heat flux at $X = 0$ was detailed in Chapter 17. Transient temperature distribution in a semi-infinite solid which is suddenly exposed to a constant heat flux $q_{X=0}$ is shown in Equation 21-1.

$$T(X,\theta)-T_i = \frac{q_{X=0}}{k}\left[\left(2\sqrt{\frac{\alpha\theta}{\pi}}\right)e^{\left(-\frac{X^2}{4\alpha\theta}\right)} - (X)\,erfc\left(\frac{X}{2\sqrt{\alpha\theta}}\right)\right] \qquad 21\text{-}1$$

A desert's surface temperature at $X = 0$ as a function of night time duration, θ, can be deduced from Equation 21-1 and it is given in Equation 21-2.

$$T(0,\theta)-T_i = \frac{2q_{X=0}}{k}\sqrt{\frac{\alpha\theta}{\pi}} \qquad 21\text{-}2$$

where $q_{X=0}$ is the sum of outgoing and incoming constant infrared radiation heat flux in W/m^2. T_i is the initial constant surface temperature of the desert at the beginning of the night in C. θ is the night time duration in seconds. k is the constant thermal conductivity of the semi-infinite desert in W/(m-K). α is the constant thermal diffusivity, the ratio of the conduction thermal energy in the semi-infinite solid to the stored thermal energy, of the semi-infinite desert in m^2/s.

$q_{X=0}$ is the constant infrared radiation heat flux at the desert's surface at night and it is composed of direct infrared radiation from its surface into space minus the downward infrared radiation from a clear sky to desert's surface or minus the downward infrared radiation from a cloudy sky to desert's surface as shown in Equations 21-3.

$$q_{X=0} = \varepsilon\alpha\left[T_i - T_{space}\right] - q_{clear\ sky} \quad \text{21-3a}$$

$$q_{X=0} = \varepsilon\alpha\left[T_i - T_{space}\right] - q_{cloudy\ sky} \quad \text{21-3b}$$

$$q_{clear\ sky} = 8.78\times10^{-13}\times T_i^{5.852}\times RH^{0.07195} \quad \text{21-3c}$$

$$q_{cloudy\ sky} = (1+K)\times8.78\times10^{-13}\times T_i^{5.852}\times RH^{0.07195} \quad \text{21-3d}$$

where Equations 21-3c and 21-3d are empirical relationships, see Ref. [11]. ε is the emissivity of desert's diffuse and gray surface in the infrared radiation region and ε is independent of direction and wavelength. σ is the Stefan-Boltzmann constant at 5.67×10^{-8} W/(m^2-K^4). T_{space} = 3 *K*. *RH* is the average relative humidity in the atmosphere in percentage. *K* is the cloudiness factor at different cloud heights.

In order to solve Equations 21-2 and 21-3, following thermophysical properties are used.

$T_i = 20C$; $\varepsilon = 0.95$

$k = 1.5\ W/(m-K)$; $\alpha = 2.52\times10^{-7}\ m^2/s$

$RH = 1\ \%$; $K = 0.34\ \textit{for cloud height} < 2\,km$

Assume a constant infrared radiation heat flux at night from desert's surface to space:

$$q_{X=0\ to\ space} = \varepsilon\alpha\left[T_i - T_{space}\right] = 0.95\times5.67\times10^{-7}\left(293^4 - 3^4\right) = \mathbf{397\ \ W/m^2}$$

Assume constant downward infrared radiation heat flux at night from a clear sky to desert's surface:

$$q_{clear\ sky} = 8.78\times10^{-13}\times T_i^{5.852}\times RH^{0.07195} = 8.78\times10^{-13}\times293^{5.852}\times1^{0.07195} = \mathbf{241\ \ W/m^2}$$

Assume constant downward infrared radiation heat flux at night from a cloudy sky to desert's surface where the cloud height is less than 2 km:

$$\begin{aligned} q_{cloudy\ sky} &= (1+K)\times8.78\times10^{-13}\times T_i^{5.852}\times RH^{0.07195} \\ &= (1+0.34)\times8.78\times10^{-13}\times293^{5.852}\times1^{0.07195} = \mathbf{323}\ W/m^2 \end{aligned}$$

Case 1: With only constant infrared radiation heat flux from desert's surface to space at night, Equation 21-2 becomes

$$T(0,\theta)=20-\frac{2\times(397-241)}{1.5}\sqrt{\frac{2.52\times10^{-7}\times\theta}{\pi}}=20-0.15\sqrt{\theta}$$

where θ is in seconds.

Case 2: With constant infrared radiation heat flux from desert's surface to space minus the constant downward infrared radiation from a clear sky to desert's surface at night, Equation 21-2 becomes

$$T(0,\theta)=20-\frac{2\times397}{1.5}\sqrt{\frac{2.52\times10^{-7}\times\theta}{\pi}}=20-0.059\sqrt{\theta}$$

where θ is in seconds.

Case 3: With constant infrared radiation heat flux from desert's surface to space minus the downward infrared radiation from a cloudy sky where the cloud height is less than 2 km to desert's surface at night, Equation 21-2 becomes

$$T(0,\theta)=20-\frac{2\times(397-323)}{1.5}\sqrt{\frac{2.52\times10^{-7}\times\theta}{\pi}}=20-0.023\sqrt{\theta}$$

where θ is in seconds.

Desert's surface temperatures at the end of a night with different night time durations are shown in Figure 21-1 for three cases given above. If there is no

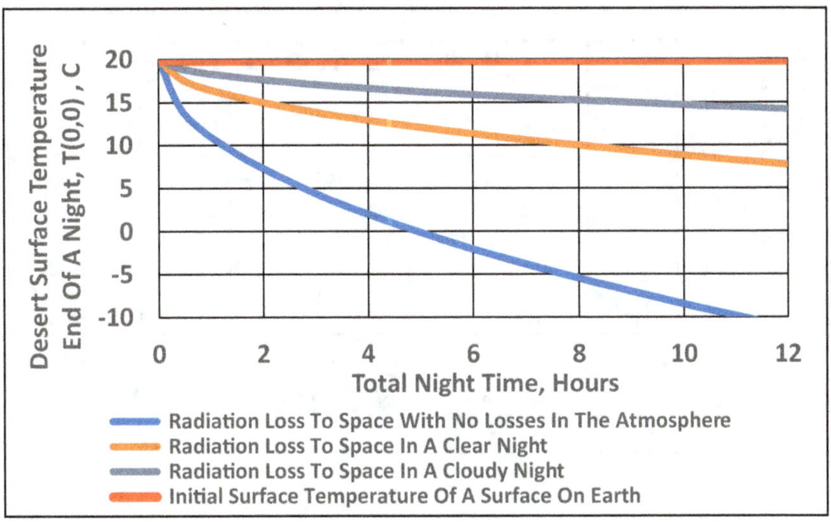

21-1 Desert surface temperatures versus night time durations in hours for three different cases given above

downward infrared radiation from the sky to desert's surface, desert's surface will freeze after a 5-hour night. If there is downward infrared radiation from a clear sky to desert's surface, desert's surface will cool down from 20 C to 7.8 C after a 12-hour night. If there is downward infrared radiation from a cloudy sky where the cloud height is less than 2 km to desert's surface, desert's surface will cool down from 20 C to 14.3 C after a 12-hour night.

Thermal conductivity of desert soil depends on water content, soil composition, porosity and soil compactness. Experimental results show that desert soil thermal conductivity can vary from 0.8 to 2.2 W/(m-K) depending on its water content and its porosity. See Ref. [4]. Figure 21-2 shows effects of desert soil's thermal conductivity for three cases in a 10-hour night.

As desert soil's thermal conductivity increases, infrared radiation heat transfer loss decreases and therefore, the desert's surface temperature loss decreases. When $k = 2.2\ W/(m - K)$, after a 10-hour night, desert's surface temperature drops to 0.6 C from 20 C for Case 1. When $k = 2.2\ W/(m - K)$, after a 10-hour night, desert's surface temperature drops to 12.4 C from 20 C for Case 2. When $k = 2.2\ W/(m - K)$, after a 10-hour night, desert's surface temperature drops only to 16.4 C from 20 C for Case 3.

Figure 21-3 shows sensitivity of desert's surface temperature to desert soil's thermal conductivity, $\frac{\partial[T(0,36,000)]}{\partial k}$, after a 10-hour night for three cases. These sensitivities are proportional to $-\frac{1}{k^2}$. Effect of increasing thermal

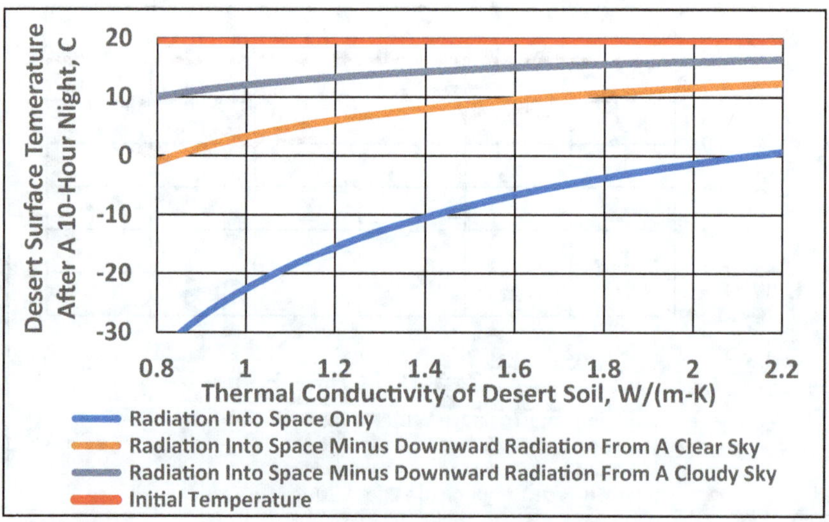

Figure 21-2 Desert surface temperature after a 10-hour night versus desert soil's thermal conductivity for three cases

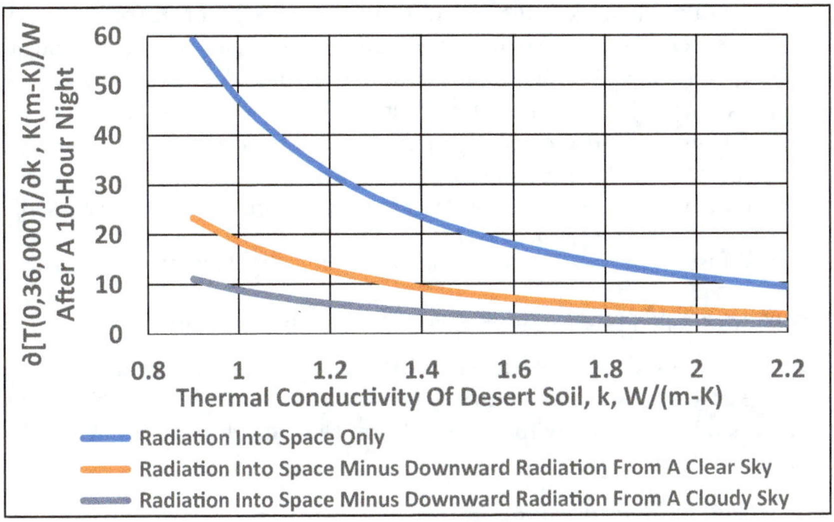

Figure 21-3 Sensitivity of desert's surface temperature to desert soil's thermal conductivity, *k*, after a 10-hour night

conductivity on desert's surface temperature decreases fast as $k = 2.2\ W/(m-K)$ value is approached. As expected, changing thermal conductivity values affect high constant surface heat flux, namely Case 1, the most.

Thermal diffusivity, namely ratio of conduction of thermal energy to storage of thermal energy, of desert soil depends mainly on water content. Experimental

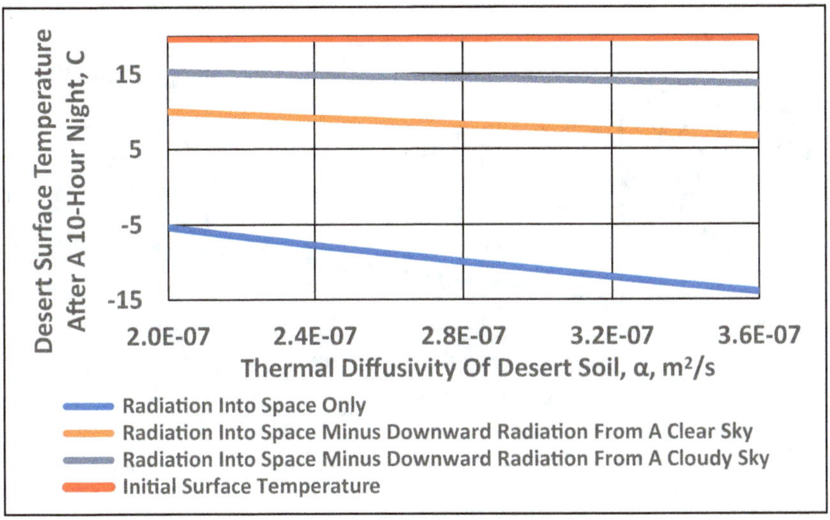

Figure 21-4 Desert surface temperature after a 10-hour night versus desert soil's thermal diffusivity for three cases

results show that desert soil thermal diffusivity can vary from 2.0×10^{-7} in dry months to 3.6×10^{-7} m^2/s on wet months of the year depending on water content. See Ref. [8]. Figure 21-4 shows effects of desert soil's thermal diffusivity for three cases in a 10-hour night. As the desert soil's thermal diffusivity increases, its surface temperature decreases due to increasing loss in net amount of infrared radiation heat flux at desert's surface.

Figure 21-5 shows sensitivity of desert's surface temperature to desert soil's thermal diffusivity, $\frac{\partial[T(0,36,000)]}{\partial\alpha}$, after a 10-hour night for three cases. These sensitivities are proportional to $\frac{1}{\alpha^{0.5}}$. Absolute sensitivity of increasing thermal diffusivity on desert's surface temperature decreases as shown in Figure 21-5.

Figure 21-6 shows the desert surface temperature after a 10-hour night versus desert air's relative humidity for three different cloud heights where in Equation 21-3d,

$$K = 0.34 \ \textit{for cloud height} < 2\,km,$$

$$K = 0.18 \ \textit{for}\ 2\,km < \textit{cloud height} < 5\,km, \text{ and}$$

$$K = 0.06 \ \textit{for cloud height} > 5\,km.$$

As the cloud height gets higher and higher, desert surface temperatures approach the clear sky temperatures. With increasing relative humidity in air, downward infrared radiation also increases and net infrared radiation on desert's

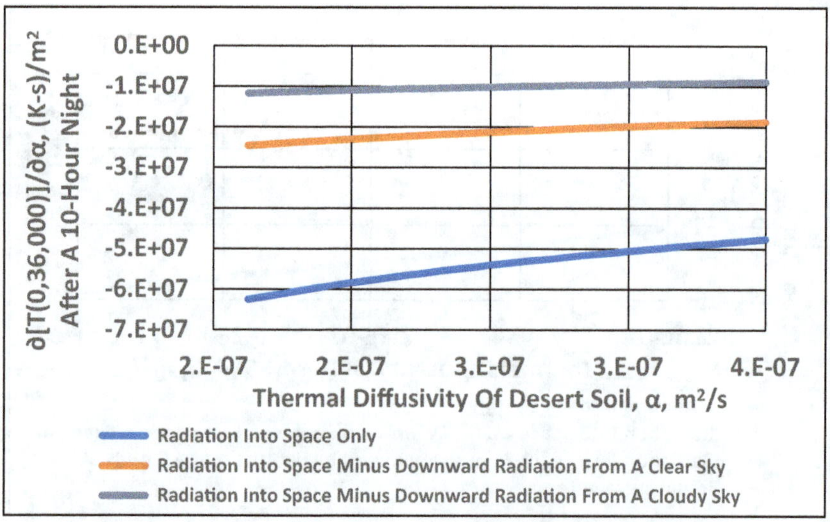

Figure 21-5 Sensitivity of desert's surface temperature to desert soil's thermal diffusivity, $\boldsymbol{\alpha}$, after a 10-hour night

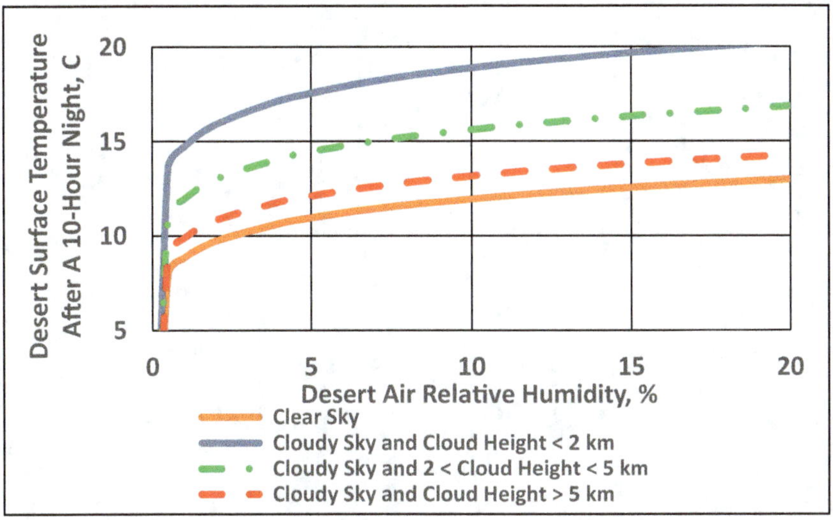

Figure 21-6 Desert surface temperature after a 10-hour night versus desert air's relative humidity

surface goes down which increases the desert's surface temperature towards the initial temperature of 20 C. However, as the relative humidity in air increases, its sensitivity to desert's surface temperature decreases, namely $\frac{\partial T(0,\theta)}{\partial(RH)} \propto \frac{1}{RH^{0.93}}$.

22

Transient Heat Transfer During Seasonal Heating and Cooling of Earth's Surface with Periodic Surface Temperature Fluctuations

Temperatures on Earth's surface fluctuates daily and seasonally in a periodic fashion that affect temperatures on its surface and in its depths. In this chapter, we will investigate transient heat transfer from a semi-infinite surface with periodic surface temperature fluctuations with respect to an average annual temperature, T_{ave}, and a maximum annual temperature, T_{max}, as given in Equations 22-1.

$$\frac{1}{\alpha}\frac{\partial T(X,\theta)}{\partial \theta} = \frac{\partial^2 T(X,\theta)}{\partial X^2} \qquad \text{22-1a}$$

$$T(0,\theta) = T_{ave} + (T_{max} - T_{ave})\cos\left[\omega\theta - (3\pi/4)\right] \qquad \text{22-1b}$$

$$T(\infty,\theta) = T_{ave} \qquad \text{22-1c}$$

In the present calculations, day 1 of the 360-day seasonal year starts at the beginning of the spring season and each season is 90 days long. Each season's average temperature is provided below.

$T_{ave-spring} = 3.33C$; $\quad T_{ave-summer} = 15.00C$

$T_{ave-fall} = 5.55C$; $\quad T_{ave-winter} = -4.44C$

360 days of annual average surface temperature is the average of four seasons' surface temperature averages given above or $T_{ave} = 4.86\ C$.

Fluctuating temperature solution to Equations 22-1 can be found in Ref. [15] and it is given in Equation 22-2.

$$T(X,\theta) = T_{ave} + (T_{max} - T_{ave})\,exp\left(-X\sqrt{\frac{\omega}{2\alpha}}\right)cos\left[\omega\theta - X\sqrt{\frac{\omega}{2\alpha}} - (3\pi/4)\right] \qquad 22\text{-}2$$

$T(X,\theta)$ is the soil temperature in C which is functions of soil depth X in meters and of time θ in seconds. ω is the angular frequency in radians per second. α is soil's constant thermal diffusivity, namely $1.4\times10^{-7}\ m^2/s$. At soil's surface, $X = 0$, Equation 22-2 reduces to soil's periodic surface temperature fluctuations or Equation 22-1b.

Only unknown in Equation 22-2 is T_{max} of soil's periodic surface temperature fluctuations. T_{max} can be obtained by equating average temperature-time area for the summer season to the fluctuating temperature-time area for the summer season as shown in Equation 22-3.

$$(15.00 - 4.86)\frac{\pi}{2} = 4.86 + (T_{max} - 4.86)\int_{90}^{180} cos\left[\frac{2\pi}{360}\theta - (3\pi/4)\right]d\theta \qquad 22\text{-}3$$

where θ is in days. Solving Equation 22-3 gives $T_{max} = 16.12\ C$. Equation 22-1b for soil's periodic surface temperature fluctuations becomes as shown in Equation 22-4 and in Figure 22-1.

$$T(0,\theta) = 4.86 + 11.14\times cos\left[\frac{2\pi}{360}\theta - (3\pi/4)\right] \qquad 22\text{-}4$$

Surface temperature seasonal periodic fluctuations in Figure 22-1 start at $\theta = 0$ *day* where $T(0,0) = 4.86 + 10.14 \times cos[-(3\pi/4)] = -3.10$ C. Surface temperature seasonal periodic fluctuations go through a maximum in the middle of the summer season when $\theta = 135$ *days* and $T_{max} = 16.12\ C$. Surface temperature seasonal periodic fluctuations goes through a minimum in the middle of the winter season when $\theta = 315$ *days* and $T_{min} = -11.26\ C$.

The skin depth, X_{sd}, in soil is defined when the argument of the exponential function in Equation 22-2 is – 1 as shown in Equation 22-5.

$$-1 = -X_{sd}\sqrt{\frac{\omega}{2\alpha}} \quad \text{or} \quad X_{sd} = \sqrt{\frac{2\alpha}{\omega}} \qquad 22\text{-}5$$

The angular frequency, ω, in radians per second is given in Equation 22-6.

$$\omega = \frac{2\pi\ radian}{360\ \ day}\times\frac{1\qquad day}{24\times3600\ second} = 2.02\times10^{-7}\,\frac{radian}{second} \qquad 22\text{-}6$$

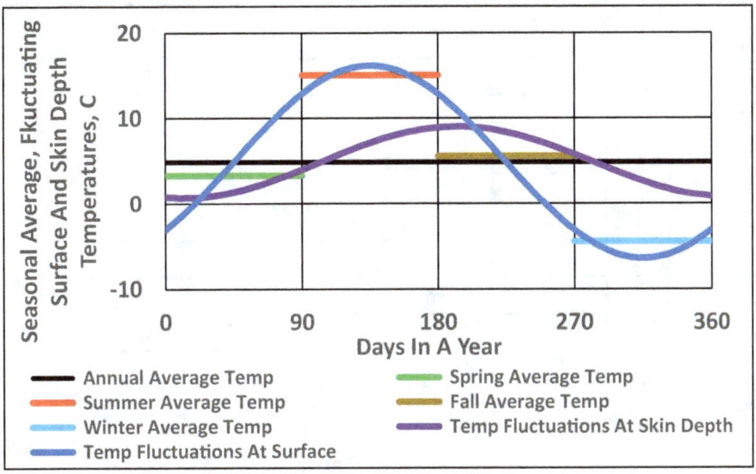

Figure 22-1 Surface and skin depth temperature fluctuations for a 360-day and a four-season year. Skin depth temperature fluctuations have a 1 radian, 57.3 degrees phase shift from surface temperature fluctuations

Then, the skin depth, X_{sd}, for the present case becomes as shown in Equation 22-7.

$$X_{sd} = \sqrt{\frac{2\alpha}{\omega}} = \sqrt{\frac{2\times1.4\times10^{-7}}{2.02\times10^{-7}}} = 1.18\,m \qquad 22\text{-}7$$

In Figure 22-1, fluctuating temperature profiles at the surface and at the skin depth of 1.18 meters can be compared. The maximum skin depth temperature drops to 9.00 C from a maximum of 16.12 C at soil's surface which can be verified as $\frac{(9.00-4.86)}{(16.12-4.86)} = 0.368 = \exp(-1)$. Similarly, the minimum skin depth-temperature increases to +0.72 C from a minimum of -9.00 C at the skin depth which can be verified as $\frac{(0.72-4.86)}{(16.12-4.86)} = -0.368 = -\exp(-1)$.

Phase shift, Φ_{sd}, at skin depth from the temperature fluctuations at the surface to the temperature fluctuations at skin depth is given in Equation 22-8.

$$\Phi_{sd} = X_{sd}\sqrt{\frac{\omega}{2\alpha}} = 1.18\sqrt{\frac{2.02\times10^{-7}}{2\times1.4\times10^{-7}}} = 1\,radian = 57.3\,degrees \qquad 22\text{-}8$$

Phase shift in temperature fluctuations versus soil depth is shown in Figure 22-2. Phase shift increases linearly with soil depth with a slope of $\sqrt{\frac{\omega}{2\alpha}}$. 360-day temperature fluctuations from soil surface to skin depth are shown using Equation

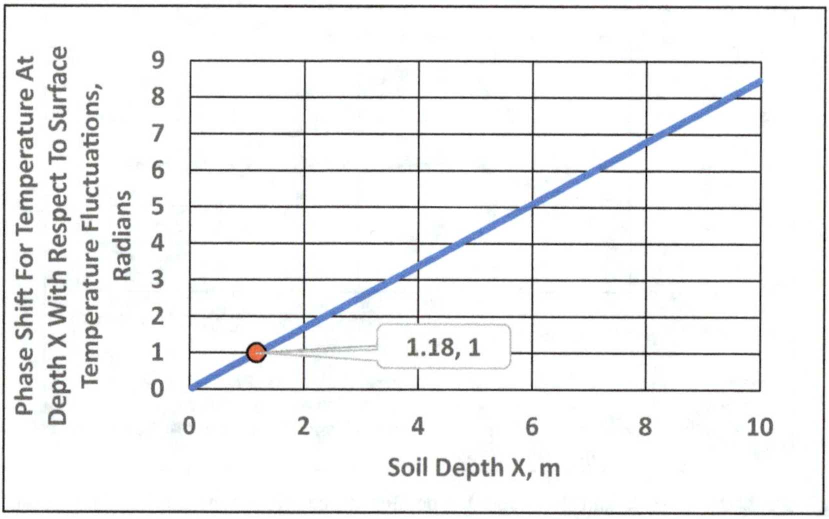

Figure 22-2 Temperature fluctuations phase shift versus soil depth X

22-2 in Figure 22-3 where $T^* = \dfrac{T(X,\theta)-T_{ave}}{T_{max}-T_{ave}}$. Two yellow dots in Figure 22-3 show the phase shift of 1 radian or 57.3 days from the maximum surface temperature in the middle of the summer season to a maximum skin depth temperature after 57.3 days in the fall season. Temperature fluctuation amplitudes decrease with soil depth as $exp\left(-X\sqrt{\dfrac{\omega}{2\alpha}}\right)$ in Equation 22-2.

Soil temperature versus soil depth using Equation 22-2 for $\theta = 0$, is shown in Figure 22-4. As soil depth increases, $T(X,0)$ asymptotes to T_{ave} which verifies the boundary condition Equation 22-1c.

Conduction heat transfer profile at soil's surface can be obtained from Equation 22-2 by differentiating it with respect to X and then setting it to $X = 0$ as shown in Equation 22-9.

$$\left(\frac{q}{A}\right)_{X=0} = -k\frac{\partial T(0,\theta)}{\partial X} = -k\left(T_{max} - T_{ave}\right)\sqrt{\frac{\omega}{\alpha}}\,cos(\omega\theta + 0.25\pi) \qquad 22\text{-}9$$

Conduction heat transfer profile at soil's surface, $\left(\dfrac{q}{A}\right)_{X=0}$, during a 360-day year is shown in Figure 22-5. Positive heat transfer is into soil surface and negative heat transfer is out of soil surface. Maximum conduction heat transfer is at $\theta = 90^o$ *or* 90 *days* at the beginning of summer season and it is +6.76 W/m^2.

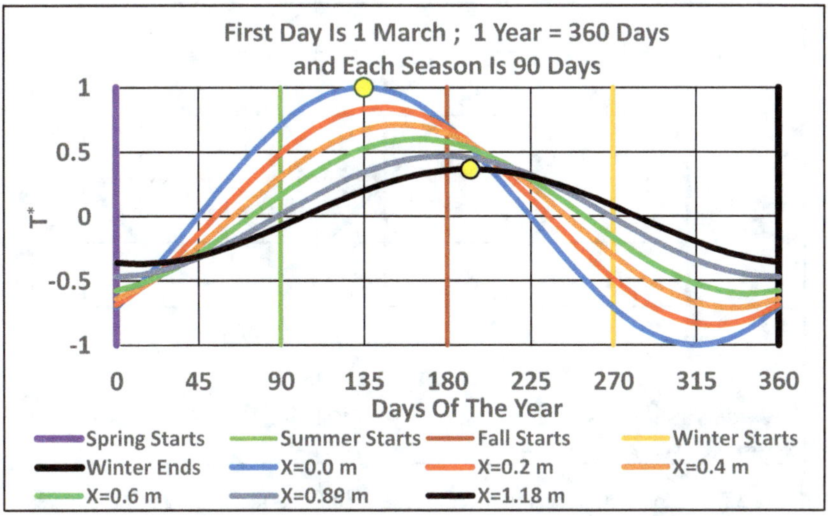

Figure 22-3 360-day temperature fluctuations from soil surface to skin depth using Equation 22-2. Two yellow dots in Figure 22-3 show the phase shift of 1 radian or 57.3 days from the maximum surface temperature fluctuation to maximum skin depth temperature fluctuation

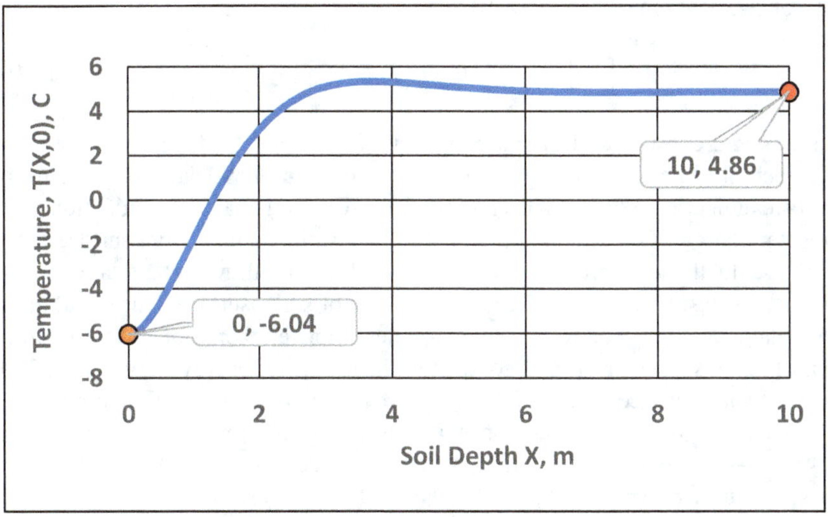

Figure 22-4 Soil temperature versus soil depth using Equation 22-2 at $\theta = 0$, beginning of spring season

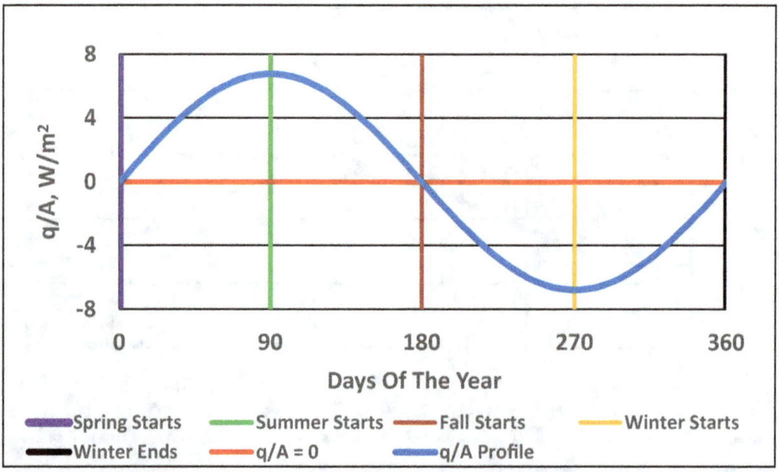

Figure 22-5 Conduction heat transfer profile at soil's surface during a 360-day year where $k = 0.5$ W/m^2. Positive heat transfer is into soil surface and negative heat transfer is out of soil surface.

Minimum conduction heat transfer is at $\theta = 270^o$ *or* 270 *days* at the beginning of winter season and it is -6.76 W/m². Total conduction heat transfer in one season can be obtained by integrating and finding the area under Equation 22-9 for one season as shown in Equation 22-10.

$$\int_0^{0.5\pi} \left(\frac{q}{A}\right)_{X=0} d\theta = k\left(T_{max} - T_{ave}\right) / \sqrt{\omega\,\alpha} \qquad 22\text{-}10$$

Assuming a constant soil thermal conductivity of $k = 0.5$ W/m^2, we get a total conduction heat transfer energy of $+3.35\times10^7$ J/m² during 90 days in spring from surface temperature fluctuations into the soil. We also get a total conduction heat transfer energy of $+3.35\times10^7$ J/m² during 90 days in summer from surface temperature fluctuations into the soil. During 90 days in fall, we get a total conduction heat transfer energy of -3.35×10^7 J/m² out of soil's surface. During 90 days in winter, we also, get a total conduction heat transfer energy of -3.35×10^7 J/m² out of soil's surface. Net total conduction heat transfer energy in one year, 360 days, at soil's surface is zero for temperature fluctuations given in Equation 22-2.

Sensitivity of total conduction heat transfer in one season at soil's surface to $(T_{max} - T_{ave})$ is linear. Sensitivity of total conduction heat transfer in 90 days at soil's surface increases by 8,88% with a 1 C increase in $(T_{max} - T_{ave})$ which shows the effect of global warming. Sensitivity of total conduction heat transfer in one season to a change in soil's thermal conductivity, *k*, or to soil's thermal capacity, ρC_p, is proportional to $\frac{1}{\sqrt{k}}$ *or* $\frac{1}{\sqrt{\rho C_p}}$. When soil's thermal conductivity, k, or soil's thermal capacity, ρC_p, increases by 10% in one season, total conduction heat transfer in one season at soil's surface increases by 4.9%. When soil's thermal conductivity, k, or soil's thermal capacity, ρC_p, decreases by 10% in one season, total conduction heat transfer in one season at soil's surface decreases by 5.1%.

23

Transient Heat Transfer from an Infinitely Long and Thin Heat Source used to Measure Thermal Conductivity of Soils and Soft Rocks in an Infinite Cylindrical Medium

Thermal conductivity of soils and soft rocks is an important thermophysical property used in models for climate change and for crop production. In this chapter, we will investigate transient heat transfer from an infinitely long and thin heat source, called the thermal needle probe, to an infinite cylindrical medium to determine the temperature rise in the medium with increasing time. We will show that the temperature-rise versus natural logarithm of time, $Ln(\theta)$, for the medium can be linearly correlated at large times. The transient heat transfer equation that applies to this case is given in radial coordinates with its initial and boundary conditions in Equations 23-1.

$$\frac{1}{\alpha}\frac{\partial T(R,\theta)}{\partial \theta}=\frac{1}{R}\frac{\partial}{\partial R}\left[R\frac{\partial T(R,\theta)}{\partial R}\right] \qquad \text{23-1a}$$

$$T(R,0)=T_i \qquad \text{23-1b}$$

$$T(\infty,\theta)=T_i \qquad \text{23-1c}$$

$$Lim_{R\to 0} k\left[2\pi R\frac{\partial T(R,\theta)}{\partial R}\right] = Q \tag{23-1d}$$

$T(R,\theta)$ is the temperature of the infinite medium in centigrade as a function of radius, R, in meters and time, θ, in seconds. T_i is the initial temperature of the medium in centigrade and as $R \to \infty$. k is the thermal conductivity of the soil or soft rock being measured in W/(m-C). Q is the heat source generated at a constant rate from a line probe at $R = 0$ which is parallel to Z-axis for $\theta > 0$ in W/m. α is the constant thermal diffusivity of the infinite medium in m^2/s.

The solution to Equations 23-1 can be obtained by using the similarity variable given in Equation 23-2.

$$\gamma = \frac{R^2}{4\alpha\theta} \tag{23-2}$$

Applying the similarity variable, we transform $\frac{\partial}{\partial R}\left[R\frac{\partial T(R,\theta)}{\partial R}\right]$ and $\frac{\partial T(R,\theta)}{\partial \theta}$ into an ordinary differential equation as shown in Equations 23-3 and 23-4.

$$\frac{\partial T}{\partial R} = \frac{dT}{d\gamma}\frac{\partial \gamma}{dR} = \frac{R}{2\alpha\theta}\frac{dT}{d\gamma} \tag{23-3a}$$

$$\frac{1}{R}\frac{\partial}{\partial R}\left[R\frac{\partial T}{\partial R}\right] = \frac{1}{R}\frac{d}{d\gamma}\left[\frac{R^2}{2\alpha\theta}\frac{dT}{d\gamma}\right]\frac{\partial \gamma}{dR} = \frac{1}{\alpha\theta}\frac{d}{d\gamma}\left(\gamma\frac{dT}{d\gamma}\right) \tag{23-3b}$$

$$\frac{\partial T}{\partial \theta} = \frac{dT}{d\gamma}\frac{\partial \gamma}{d\theta} = -\frac{R^2}{2\alpha\theta^2}\frac{dT}{d\gamma} \tag{23-3c}$$

Using the similarity variable, Equations 23-1a become as shown in Equation 23-4a. Equation 23-4b complies both with the initial condition Equation 23-1b and the boundary condition Equation 23-1c. The boundary condition in Equation 23-1d becomes Equation 23-4c.

$$-\gamma\frac{dT}{d\gamma} = \frac{d}{d\gamma}\left(\gamma\frac{dT}{d\gamma}\right) \tag{23-4a}$$

$$T(\infty) = T_i \tag{23-4b}$$

$$Lim_{\gamma\to 0}\left[\gamma\frac{dT}{d\gamma}\right] = \frac{Q}{4\pi k} \tag{23-4c}$$

Separating $\gamma \frac{dT}{d\gamma}$ in Equation 23-4a, we get Equation 23-5.

$$\frac{d\left(\gamma \frac{dT}{d\gamma}\right)}{\left(\gamma \frac{dT}{d\gamma}\right)} = -d\gamma \qquad 23\text{-}5$$

Integrating Equation 23-5, we get Equation 23-6.

$$Ln\left(\gamma \frac{dT}{d\gamma}\right) = -\gamma + C_1 \qquad 23\text{-}6$$

Equation 23-6 can be rewritten as shown in Equations 23-7a.

$$\gamma \frac{dT}{d\gamma} = C_2 e^{-\gamma} \qquad 23\text{-}7a$$

Applying the boundary condition in Equation 23-4c, C_2 becomes $\frac{Q}{4\pi k}$ and Equation 23-7a takes the form as shown in Equation 23-7b.

$$\frac{dT}{d\gamma} = \frac{Q}{4\pi k}\frac{e^{-\gamma}}{\gamma} \qquad 23\text{-}7b$$

Integrating Equation 23-7b, we get Equation 23-8.

$$T(\gamma) = \frac{Q}{4\pi k}\int_{\gamma}^{\infty} \frac{e^{-l}}{l} dl + C_3 \qquad 23\text{-}8$$

where l is a dummy variable. Applying the boundary condition in Equation 23-4b, C_3 becomes T_i and Equation 23-8 takes the final form as shown in Equation 23-9.

$$T(\gamma) - T_i = \frac{Q}{4\pi k}\int_{\gamma}^{\infty} \frac{e^{-l}}{l} dl \qquad 23\text{-}9$$

The integral $\int_{\gamma}^{\infty} \frac{e^{-l}}{l} dl$ is the exponential integral designated by $E_1(\gamma)$. The exponential integral can be expanded in series format for small γ or large time θ as shown in Equation 23-10.

$$E_1(\gamma) = -C_{E-M} - Ln(\gamma) + \gamma - \frac{\gamma^2}{4} + \frac{\gamma^3}{18} - O(\gamma^4) \qquad 23\text{-}10$$

where C_{E-M} is the Euler-Mascheroni constant which is the limiting difference between harmonic series and the natural logarithm given in Equation 23-11.

$$C_{E-M} = 0.57721566490153286060 65... \qquad 23\text{-}11$$

We can use first two terms of the exponential integral to a good approximation of $(T(\gamma) - T_i)$ for large θ as shown in Equation 23-12.

$$T(\gamma) - T_i \approx \frac{Q}{4\pi k}\left(-C_{E-M} - Ln\left(\frac{R^2}{4\alpha\theta}\right)\right) \qquad 23\text{-}12$$

We can rewrite Equation 23-12 by separating the time, θ, term in the natural-logarithm as shown in Equation 23-13.

$$T\left(\frac{R^2}{4\alpha\theta}\right) - T_i \approx \frac{Q}{4\pi k}Ln(\theta) + \frac{Q}{4\pi k}\left(-C_{E-M} - Ln\left(\frac{R^2}{4\alpha}\right)\right) \qquad 23\text{-}13$$

where the second term in Equation 23-13 is a constant since Q, R, and soil's thermal properties are assumed to be constants during temperature-time measurements with a long and thin heat source, namely the thermal needle probe. A more detailed derivation of Equation 23-13 can be found in Carslaw & Jaeger, Ref. [5].

Finally, from the linear slope of temperature-natural logarithm of time measurements, we can obtain thermal conductivity of soils and soft rocks with an accuracy of ±3% by following procedures in standards such as ASTM-5334-08 and IEEE 442-1981 with the relationship shown in Equation 23-14.

$$k = \frac{\left[Ln(\theta)_2 - Ln(\theta)_1\right]}{\left[\left(\frac{4\pi}{Q}\right)\left(\Delta T_2 - \Delta T_1\right)\right]} \qquad 23\text{-}14$$

where $\Delta T = T\left(\frac{R^2}{4\alpha\theta}\right) - T_i$.

Equation 23-14 is valid for large time values, namely, $\frac{R^2}{4\alpha\theta} << 1$. Also, water content and moisture in the soil or in the rock affect correct determination of thermal conductivity. A typical temperature-Ln(time) measurement data profile with a long and thin heat source, a thermal needle probe inserted into a dry soil or a soft rock, is shown in Figure 23-1. The constant in Equation 23-13, namely the constant lag in $Ln(\theta)$ axis, is not shown in Figure 23-1. In the below graph, the initial rise of temperature should also be neglected, since this initial rise of temperature is affected by the thermal capacity of the probe and by the contact resistance between the probe's surface and the medium. In Figure 23-1, we can consider the slope of this curve for $\theta > 11$ *s* or $Ln(\theta) > 2.4$.

Let us use the data at two red dots in the linear region in Figure 23-1 to calculate the thermal conductivity. The constant heat input, Q, from the probe is 4 W/m. When linear slope of the curve between time at 21 s and at 55 s, between two red dots, is considered,

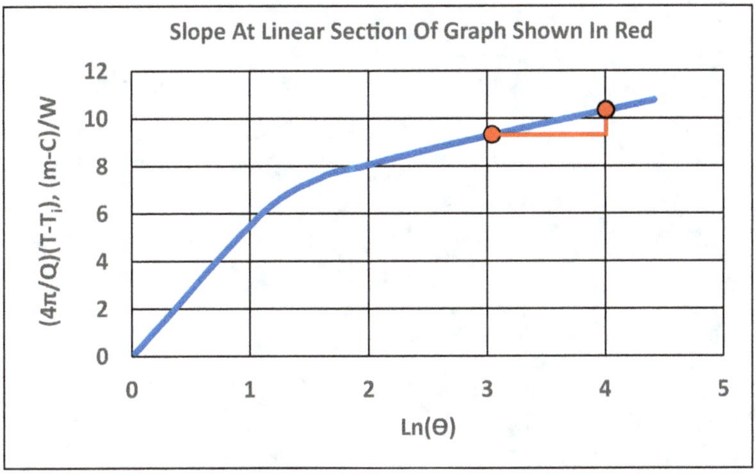

Figure 23-1 Typical temperature-time measurement data profile without the constant lag in $Ln(\theta)$ - axis

$$Ln(\theta)_2 - Ln(\theta)_1 = 0.9628$$

and

$$\left(\frac{4\pi}{Q}\right)(\Delta T_2 - \Delta T_1) = 1.0259.$$

With $Q = 4.0\ W/m$, the thermal conductivity of this soil is

$$k = \frac{\left[Ln(\theta)_2 - Ln(\theta)_1\right]}{\left(\frac{4\pi}{Q}\right)(\Delta T_2 - \Delta T_1)} = \frac{0.9628}{1.0259} = 0.94W/(m-C).$$

24

Solidification of Magmas in Lava Lakes

Erupting volcanoes flow sometimes into pit craters located around them and form lava lakes. After volcanic eruption stops, magmas in these pit craters start to solidify and release latent heat of fusion to the atmosphere from its solidified surface. It can take years to complete solidification of the liquid magma in some deep pit craters. Pit craters are not explosion craters or vents. Pit craters are depressions of a localized collapse into a void. Time dependent heat transfer during solidification of magma in a lava lake in a pit crater is similar to ice formation on a water lake and it is shown in Figure 24-1 where X is the thickness of the solidified magma layer which started solidifying from $X = 0$.

In Figure 24-1, T_{atm} is the constant temperature of the atmosphere in Kelvin above the top of the solidified magma layer where $X = 0\ m$. H_L is the constant latent heat of fusion in J/kg which is released from the phase change interface, $X_{int}(\theta)$ in meters, from liquid to solid. θ is the solidification time in seconds.

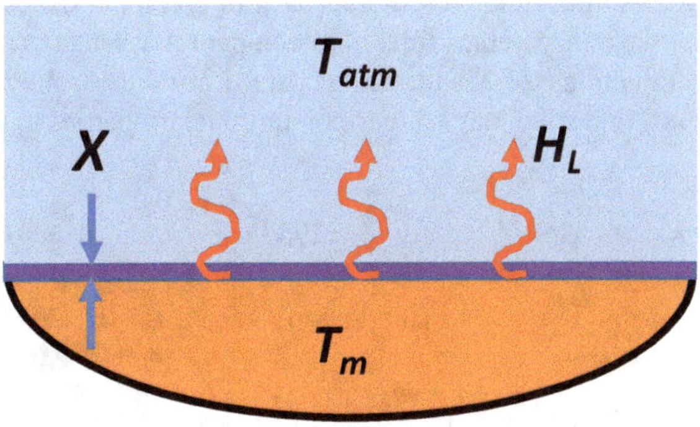

Figure 24-1 Magma layer solidification on the surface of a pit crater

T_m is the constant temperature of liquid magma everywhere in the pit crater below the phase change interface.

Latent heat of fusion released from phase change interface has to escape through the solidified magma layer and from the solidified magma surface at $X = 0$ by going through conduction and convection thermal resistances as shown in Equation 24-1.

$$\rho H_L \frac{dX}{d\theta} = \frac{(T_m - T_{atm})}{\frac{1}{h_{atm}} + \frac{X}{k_s}} \qquad 24\text{-}1$$

where ρ is the density of the solidified magma in kg/m^3. k_s is the thermal conductivity of solidified magma in W/(m-K). h_{atm} is the convection plus radiation heat transfer coefficient between the solidified magma surface at $X = 0$ and the atmosphere. The initial condition for the above first order differential equation is shown in Equation 24-2.

$$X = 0 \text{ at } \theta = 0 \qquad 24\text{-}2$$

The solution to Equations 24-1 and 24-2 is a second order polynomial in X and it is shown in Equations 24-3.

$$X(\theta) = \frac{-B + \sqrt{B^2 + 4AC}}{2A} \qquad 24\text{-}3a$$

$$A = 1; \quad B = \frac{2k_s}{h_{atm}}; \quad C = -2\alpha\left[\frac{C_P(T_m - T_{atm})}{H_L}\right]\theta \qquad 24\text{-}3b$$

Thermophysical properties of liquid, solidifying, and solid magma depend on temperature-time histories and on rheological-time histories which are very complex. Best way to simplify the present approach to predict thickness of solidifying crust in a pit lava lake versus time is to compare X versus θ results that are obtained from Equations 24-3 with actual data. Let us use following constant liquid, solidifying, and solid magma thermos-physical properties to solve Equations 24-3.

$$T_m = 1{,}050\,K; \qquad T_{atm} = 293\,K; \qquad H_L = 410{,}000\,J/kg$$

$$\rho = 900\ kg/m^3; \qquad C_p = 900\,J/(kg - K) \quad k_s = 0.5\ W/(m - K)$$

$$\alpha = \frac{k_s}{\rho C_p} = 6.173 \times 10^{-7}\ m^2/s; \quad h_{atm} = 20\ W/(m^2 - K)$$

Solidified magma thickness versus time comparisons of Equation 24-3 results using above constant thermo-physical properties and actual measurements of solidified magma thicknesses on three pit lava lakes in Kilauea, Hawaii, from Ref. [16], are shown in Figure 24-2. Actual measurements of solidified magma thicknesses versus time on three pit lava lakes are shown as colored circles.

The parabolic growth of solidified magma thickness fits measured data very well. Now, let us investigate sensitivity of solidified magma thickness at a given time, $\frac{dX}{d\theta}$, namely after 12 months and after 48 months, to ±10% variations in constant thermo-physical properties given above. $C_P\,(T_m - T_{atm})/\,H_L$ sensitivity results are shown below in Table 24-1.

$C_P\,(T_m - T_{atm})/\,H_L$ is the ratio of the thermal energy that a solidified magma layer can store to latent heat of fusion released from the solidifying interface. If the thermal energy storage capability of the solidified magma layer decreases by 10%, solidified magma layer thickness decreases by 5.15% after 12 months and it decreases by 5.14% after 48 months. On the other hand, If the thermal energy storage capability of the solidified magma layer increases by 10%, solidified magma layer thickness increases by 4.90% after 12 months and it increases by 4.89% after 48 months. Changes in latent heat of fusion, H_L, reverses the sensitivity of change in X to change in time θ. When H_L decreases by 10%, X increases by 5.43% after 12 months and increases by 5.42% after 48 months. When H_L increases by 10%, X decreases by 4.67% after 12 months and decreases by 4.66% after 48 months.

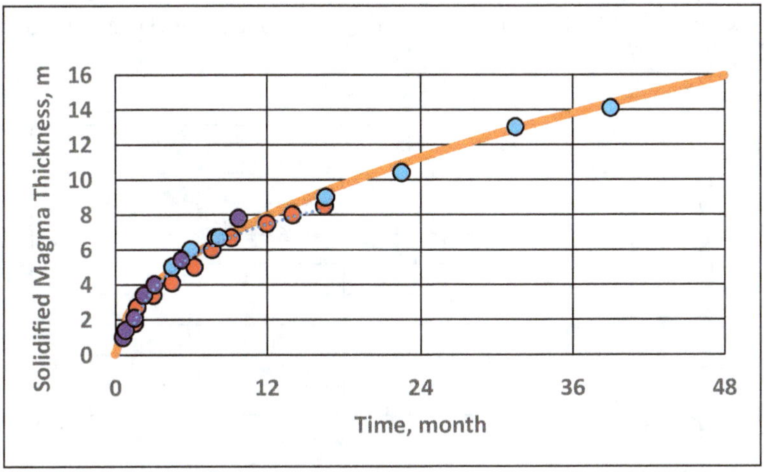

Figure 24-2 Solidified magma thickness versus time comparisons of Equation 24-3 results using above constant thermo-physical properties shown in orange colored line and actual solidified magma thickness versus time measurements are shown in colored circles from three pit lava lakes in Kilauea, Hawaii, Ref. [16].

Non-Dimensional Thermo-Physical Property $C_P (T_m - T_{atm})/H_L$	Nominal	-10%	+10%
$C_P (T_m - T_{atm})/H_L$	1.662	1.496	1.828
X in m. in 12 mo.	7.963	7.553	8.353
% Change in Solidified Thickness		-5.15%	+4.90 %
X in m. in 48 mo.	15.951	15.131	16.731
% Change in Solidified Thickness		-5.14 %	+4.89 %

Table 24-1 Sensitivity of solidified magma thickness, ***X***, at 12 months and at 48 months to ±10% change in $\boldsymbol{C_P (T_m - T_{atm})/ H_L}$

Thermal diffusivity, α, is solidified magma's ability to conduct latent heat of fusion relative to its ability to store thermal energy. Thermal diffusivity's sensitivity to ±10% change in its value is the same as the ±10% change in $C_P (T_m - T_{atm})/ H_L$ value.

Constant thermal conductivity, k_s, is solidified magma's ability to conduct latent heat of fusion to the $X = 0$ surface. The temperature in the solid magma layer varies linearly with X. When k_s decreases by 10%, X decreases by 5.12% after 12 months and decreases by 5.12% after 48 months. When k_s increases by 10%, X increases by 4.86% after 12 months and increases by 4.87% after 48 months. Figure 24-3 shows the solidified magma thickness versus time for the nominal k_s and for ±10% k_s.

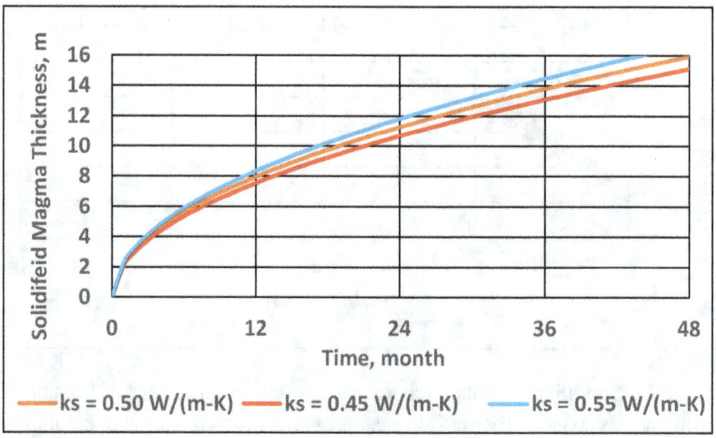

Figure 24-3 Solidified magma thickness versus time for the nominal k_s and for ±10% k_s

Constant convection plus radiation heat transfer coefficient h_{atm} between the solidified surface at $X = 0$ and the atmosphere creates another thermal resistance, $\frac{1}{h_{atm}}$, to latent heat of fusion to escape. When h_{atm} decreases by 10%, X decreases by 0.035% after 12 months and decreases by 0.017% after 48 months. When h_{atm} increases by 10%, X increases by 0.028% after 12 months and increases by 0.014% after 48 months.

Changes in convection plus radiation heat transfer coefficient h_{atm} between the solidified surface at $X = 0$ and the atmosphere have little effect on solidification layer of magma except during initial stages of solidification. This effect is shown during first day of solidification in Figure 24-4. A 10% reduction in h_{atm}, reduces the solidified magma layer thickness after a day by 4.49%. A 10%, increase in h_{atm}, increases the solidified magma layer thickness after a day by 3.90%. As a comparison, a 10% reduction in h_{atm}, reduces the solidified magma layer thickness after a year by 0.035%. A 10%, increase in h_{atm}, increases the solidified magma layer thickness after a year by 0.028%.

Since the convection plus radiation heat transfer coefficient h_{atm} has minor effect on X versus θ profiles or the release of latent heat of fusion from $X = 0$ surface, we can eliminate h_{atm} from Equation 24-1 by letting $h_{atm} \to \infty$ or by letting its thermal resistance go to zero as shown in Equation 24-4.

$$\rho H_L \frac{dX}{d\theta} = \frac{k_s \left(T_m - T_{atm}\right)}{X} \tag{24-4}$$

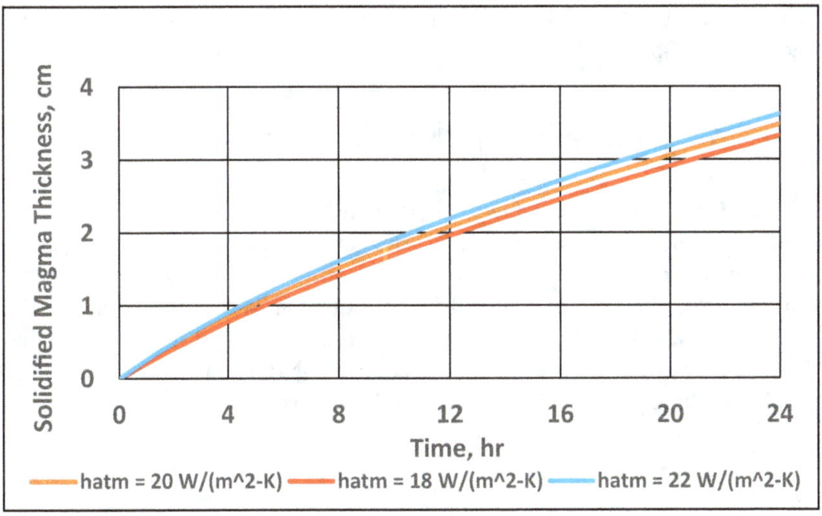

Figure 24-4 Solidified magma thickness versus time for the nominal h_{atm} and for ±10% h_{atm} during first day of solidification

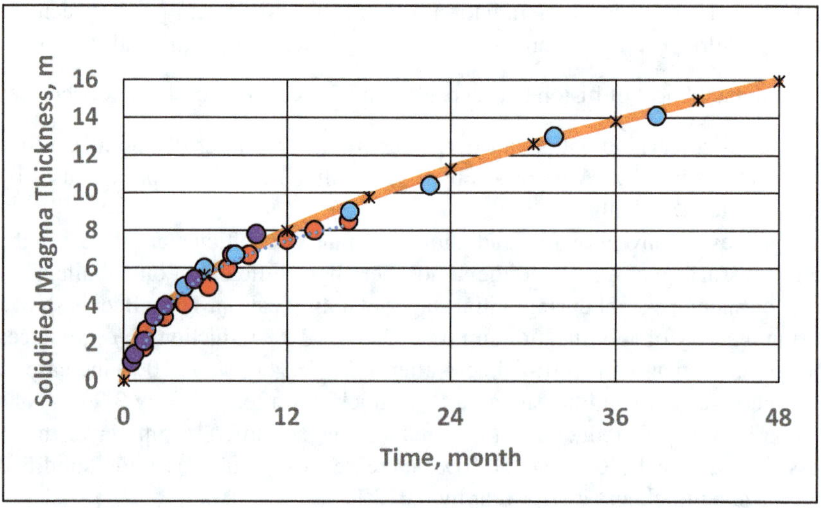

Figure 24-5 Solidification magma thickness versus time comparisons of Equation 24-5 results using above constant thermo-physical properties and actual solidified magma thickness versus time measurements shown in colored circles from three pit lava lakes in Kilauea, Hawaii, from Ref. [16]. "x" values are shown every six months resulting from Equations 24-3 for comparison purposes to orange colored line resulting from Equation 24-5.

The solution to Equation 24-4 and the initial condition at Equation 24-2 is shown in Equations 24-5.

$$X(\theta) = \sqrt{2\alpha\left[\frac{C_P(T_m - T_{atm})\theta}{H_L}\right]} \qquad 24\text{-}5$$

With same constant thermos-physical properties given above, we can plot the solidified magma layer thickness versus time and actual measurements of solidified magma thicknesses versus time on three pit lava lakes in Kilauea, Hawaii, from Ref. [16], in Figure 24-5. Results from Equation 24-5 shown in orange line and Equations 24-3 results are shown in "x". They are almost the same and they follow measured solidified magma thickness versus time data shown in colored circles from three pit lava lakes in Kilauea, Hawaii very closely.

References

1. Atesmen, M.K. (2009). *Everyday Eat Transfer Problems – Sensitivities to Governing Variables*. New York: ASME Press.
2. Atesmen, M.K. (2021). *Electromagnetic Waves and Heat Transfer – Sensitivities to Governing Variable in Everyday Life*. New York: ASME Press.
3. Atesmen, M.K. (2019). Case Studies in Fluid Mechanics with *Sensitivities to Governing Variables*. Hoboken, New Jersey: John Wiley & Sons Ltd. and New York: ASME Press.
4. Atwani, H. and et al. (2018). *Evaluating Thermal Conductivity of Desert Sand Under Different Initial Properties*. JATER, V4, p. 236-240.
5. Carslaw, H.S. and Jaeger, J.C. (1959). *Conduction of Heat in Solids*. 2e. Oxford University Press, Oxford.
6. Dergarabedian, P. (1960). *Observations on Bubble Growth in Various Superheated Liquids*. J. of Fluid Mechanics, V9, p. 39-48.
7. Galante, M, et al. (2010). *Eggstraordinary Eggsperiments – An Activity to Teach Heat Transfer Concepts to K-12 Students*. University of Michigan, Ann Arbor.
8. Gao, Z. and et al. (2017). *Determination of Desert Soil Apparent Thermal Conductivity Using a Conduction-Convection Algorithm*. JGR. P. 9560-9577.
9. Holman, J.P. (1968). *Heat Transfer*. New York, McGraw-Hill Book Company.
10. Incropera, F.P. and DeWitt, D.P. (1996). *Fundamentals of Heat and Mass Transfer*. 4e. New York: John Wiley & Sons Ltd.
11. Maasoumy, M. and et al. (2013) *Online Simultaneous State Estimation and Parameter Adaptation for Building Predictive Control*. University of California, Berkeley.
12. NOAA (National Oceanic and Atmospheric Administration). (2015). National Solar Radiation Database.
13. Robinson, A.J. (2002). *Bubble Growth Dynamics in Boiling*. Ph. D. Thesis, McMaster University, Hamilton, Ontario, Canada
14. Rohsenow, W.M. and Choi, H.Y. (1961). *Heat, Mass and Momentum Transfer*. Englewood Cliff, New Jersey: Prentice-Hall, Inc.

15. Schneider, P.J. (1955). *Conduction Heat Transfer*. Reading, Massachusetts: Addison-Wesley Publishing Company, Inc.
16. Turcotte, D.L. and Schubert, G. (2002). *Geodynamics*. 2e. Cambridge University Press, Cambridge, UK.
17. Wardowski, W.F. et al. (1997). *Separation and Grading of Freeze-Damaged Citrus Fruits*. University of Florida, Gainsville.

Appendix

Derivation of Heat Conduction Equations in Cartesian, Cylindrical and Spherical Coordinates

In this appendix, we shall derive differential equations governing temperature distribution in a body, namely the heat conduction equation, in Cartesian, cylindrical and spherical coordinates. First, let us derive the general heat conduction equation in Cartesian coordinates using a differential rectangular parallelepiped body with sides dX, dY, and dZ as shown in Figure A-1.

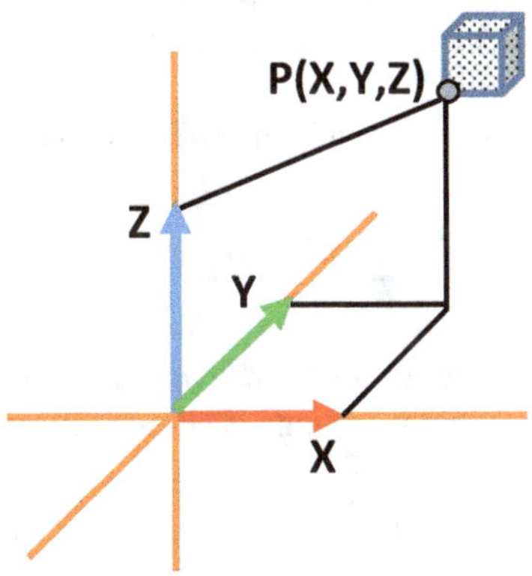

Figure A-1 A differential rectangular parallelepiped in Cartesian coordinates with sides dX, dY, and dZ that are parallel to X, Y, and Z, respectively. The location of the closest corner of the rectangular parallelepiped body to the origin of the Cartesian coordinates is designated by P(X,Y,Z).

Using the first law of thermodynamics, energy balance for conduction heat flow for the differential rectangular parallelepiped during a small time $d\theta$ is given by Equation A-1.

$$\textit{Heat flow in} - \textit{Heat flow out} + \textit{Thermal energy generation} = \textit{Thermal energy change} \qquad \text{A-1}$$

Using Fourier's Law for heat conduction, let us write the energy balance for the differential rectangular parallelepiped for conduction heat flow in and conduction heat flow out in the X-direction as shown in Equation A-2.

$$\left(-k\frac{\partial T}{\partial X}\right)dYdZ - \left[-k\frac{\partial T}{\partial X} + \frac{\partial}{\partial X}\left(-k\frac{\partial T}{\partial X}\right)dX\right]dYdZ = \frac{\partial}{\partial X}\left(k\frac{\partial T}{\partial X}\right)dXdYdZ \qquad \text{A-2}$$

In Equation A-2, the conduction heat flow out term uses only first two terms of the Taylor series expansion for the change in conduction heat transfer from location X to location X+dX.

Similar energy balances for the differential rectangular parallelepiped for conduction heat flow in and conduction heat flow out in the Y and in the Z-directions are shown in Equations A-3 and A-4.

$$\left(-k\frac{\partial T}{\partial Y}\right)dXdZ - \left[-k\frac{\partial T}{\partial Y} + \frac{\partial}{\partial Y}\left(-k\frac{\partial T}{\partial Y}\right)dY\right]dXdZ = \frac{\partial}{\partial Y}\left(k\frac{\partial T}{\partial Y}\right)dXdYdZ \qquad \text{A-3}$$

$$\left(-k\frac{\partial T}{\partial Z}\right)dXdY - \left[-k\frac{\partial T}{\partial Y} + \frac{\partial}{\partial Z}\left(-k\frac{\partial T}{\partial Z}\right)dZ\right]dXdY = \frac{\partial}{\partial Z}\left(k\frac{\partial T}{\partial Z}\right)dXdYdZ \qquad \text{A-4}$$

k is the thermal conductivity of the differential rectangular parallelepiped body in W/(m-K). T is the temperature of the differential rectangular parallelepiped body in K. X, Y, and Z are length dimensions in Cartesian coordinates in meters. θ is time in seconds.

Rate of thermal energy generation in the differential rectangular parallelepiped body is given in Equation A-5.

$$\dot{q}dXdYdZ \qquad \text{A-5}$$

where $\dot{q}$ is in J/(m^3-s).

Thermal energy change per unit time $d\theta$ in the differential rectangular parallelepiped body without any phase change is shown in Equation A-6.

$$\rho C_p \frac{\partial T}{\partial \theta} dXdYdZ \qquad \text{A-6}$$

where ρ is the density for the differential rectangular parallelepiped body in kg/m^3. C_p is the specific heat at constant pressure for the differential rectangular parallelepiped body in J/(kg-K).

Inserting Equations A-2 through A-6 into A-1 and dividing by differential rectangular parallelepiped body's volume, $dXdYdZ$, we get the following second order partial differential equation for heat conduction in Cartesian coordinates as shown in Equation A-7.

$$\frac{\partial}{\partial X}\left(k\frac{\partial T}{\partial X}\right)+\frac{\partial}{\partial Y}\left(k\frac{\partial T}{\partial Y}\right)+\frac{\partial}{\partial Z}\left(k\frac{\partial T}{\partial Z}\right)+\dot{q}=\rho C_p\frac{\partial T}{\partial \theta} \qquad \text{A-7}$$

For constant thermal conductivity, Equation A-7 becomes as shown in Equation A-8.

$$\frac{\partial^2 T}{\partial X^2}+\frac{\partial^2 T}{\partial Y^2}+\frac{\partial^2 T}{\partial Z^2}+\frac{\dot{q}}{k}=\frac{1}{\alpha}\frac{\partial T}{\partial \theta} \qquad \text{A-8}$$

where $\alpha = \frac{k}{\rho C_p}$ is the thermal diffusivity of body's material in m^2/s. Thermal diffusivity represents the ratio of conduction thermal energy through that body material to storage capacity of thermal energy in that body material.

Now, let us change the second order partial differential Equation A-8 for heat conduction in Cartesian coordinates to heat conduction in cylindrical coordinates as shown in Figure A-2.

Cartesian coordinates are represented by cylindrical coordinates in Equation A-9.

$$X = rcos\varphi \quad ; Y = rsin\varphi \quad ; Z = Z \qquad \text{A-9}$$

Partial derivatives of **X,Y,Z** with respect to **r,Φ,Z** are given in Equations A-10.

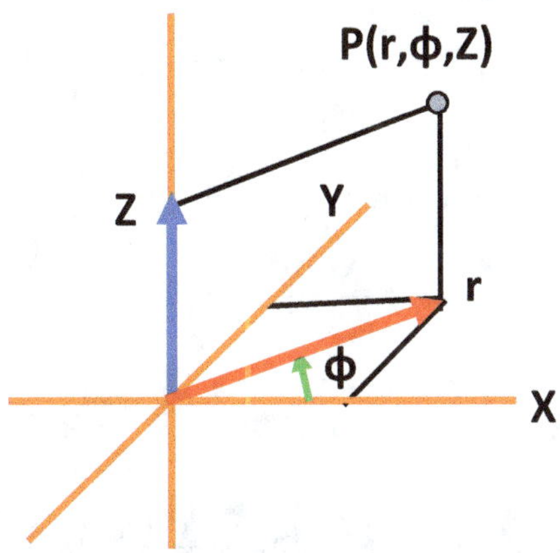

Figure A-2 Cartesian coordinates X,Y,Z and cylindrical coordinates r,Φ,Z . Point P(r,Φ,Z) is defined in cylindrical coordinates

$$\frac{\partial X}{\partial r} = cos\varphi;\ \frac{\partial X}{\partial \varphi} = -rsin\varphi;\frac{\partial X}{\partial Z} = 0 \qquad \text{A-10a}$$

$$\frac{\partial Y}{\partial r} = sin\varphi;\ \frac{\partial Y}{\partial \varphi} = rcos\varphi;\frac{\partial Y}{\partial Z} = 0 \qquad \text{A-10b}$$

$$\frac{\partial Z}{\partial r} = 0;\frac{\partial Z}{\partial \varphi} = 0;\frac{\partial Z}{\partial Z} = 1 \qquad \text{A-10c}$$

We can now use the **chain rule** of differentiating because (X,Y,Z) are functions of (r,φ,Z) in order to obtain $\frac{\partial T}{\partial r},\frac{\partial T}{\partial \varphi},\frac{\partial T}{\partial Z},\frac{\partial^2 T}{\partial r^2},\frac{\partial^2 T}{\partial \varphi^2},\frac{\partial^2 T}{\partial Z^2}$ in cylindrical coordinates as functions of temperature derivatives in Cartesian coordinates.

$\frac{\partial T}{\partial r}$ is shown in Equation A-11.

$$\frac{\partial T}{\partial r} = \frac{\partial T}{\partial X}\frac{\partial X}{\partial r} + \frac{\partial T}{\partial Y}\frac{\partial Y}{\partial r} + \frac{\partial T}{\partial Z}\frac{\partial Z}{\partial r} = cos\varphi\frac{\partial T}{\partial X} + sin\varphi\frac{\partial T}{\partial Y} + 0 \qquad \text{A-11}$$

$\frac{\partial T}{\partial \varphi}$ is shown in Equation A-12.

$$\frac{\partial T}{\partial \varphi} = \frac{\partial T}{\partial X}\frac{\partial X}{\partial \varphi} + \frac{\partial T}{\partial Y}\frac{\partial Y}{\partial \varphi} + \frac{\partial T}{\partial Z}\frac{\partial Z}{\partial \varphi} = -rsin\varphi\frac{\partial T}{\partial X} + rcos\varphi\frac{\partial T}{\partial Y} + 0 \qquad \text{A-12}$$

$\frac{\partial T}{\partial Z}$ is shown in Equation A-13.

$$\frac{\partial T}{\partial Z} = \frac{\partial T}{\partial X}\frac{\partial X}{\partial Z} + \frac{\partial T}{\partial Y}\frac{\partial Y}{\partial Z} + \frac{\partial T}{\partial Z}\frac{\partial Z}{\partial Z} = 0 + 0 + \frac{\partial T}{\partial Z} \qquad \text{A-13}$$

$\frac{\partial^2 T}{\partial r^2}$ is derived in Equations A-14.

$$\frac{\partial^2 T}{\partial r^2} = cos\varphi\frac{\partial}{\partial r}\left(\frac{\partial T}{\partial X}\right) + sin\varphi\frac{\partial}{\partial r}\left(\frac{\partial T}{\partial Y}\right) \qquad \text{A-14a}$$

$$\frac{\partial^2 T}{\partial r^2} = cos\varphi\left[\frac{\partial}{\partial X}\left(\frac{\partial T}{\partial X}\right)\left(\frac{\partial X}{\partial r}\right) + \frac{\partial}{\partial Y}\left(\frac{\partial T}{\partial X}\right)\left(\frac{\partial Y}{\partial r}\right)\right] + sin\varphi\left[\frac{\partial}{\partial X}\left(\frac{\partial T}{\partial Y}\right)\left(\frac{\partial X}{\partial r}\right) + \frac{\partial}{\partial Y}\left(\frac{\partial T}{\partial Y}\right)\left(\frac{\partial Y}{\partial r}\right)\right] \qquad \text{A-14b}$$

$$\frac{\partial^2 T}{\partial r^2} = cos^2\varphi\frac{\partial^2 T}{\partial X^2} + 2cos\varphi sin\varphi\frac{\partial^2 T}{\partial X \partial Y} + sin^2\varphi\frac{\partial^2 T}{\partial Y^2} \qquad \text{A-14c}$$

$\frac{\partial^2 T}{\partial \varphi^2}$ is derived in Equations A-15.

$$\frac{\partial^2 T}{\partial \varphi^2} = -rcos\varphi\frac{\partial T}{\partial X} - rsin\varphi\left[\frac{\partial}{\partial X}\left(\frac{\partial X}{\partial \varphi}\right)\left(\frac{\partial T}{\partial X}\right) + \frac{\partial}{\partial Y}\left(\frac{\partial Y}{\partial \varphi}\right)\left(\frac{\partial T}{\partial X}\right)\right] - rsin\varphi\left(\frac{\partial T}{\partial Y}\right) + rcos\varphi\left[\frac{\partial}{\partial X}\left(\frac{\partial X}{\partial \varphi}\right)\left(\frac{\partial T}{\partial Y}\right) + \frac{\partial}{\partial Y}\left(\frac{\partial Y}{\partial \varphi}\right)\left(\frac{\partial T}{\partial Y}\right)\right] \quad \text{A-15a}$$

$$\frac{\partial^2 T}{\partial \varphi^2} = -r\left(cos\varphi\frac{\partial T}{\partial X} + sin\varphi\frac{\partial T}{\partial Y}\right) + r^2\left(sin^2\varphi\frac{\partial^2 T}{\partial X^2} - 2cos\varphi sin\varphi\frac{\partial^2 T}{\partial X \partial Y} + cos^2\varphi\frac{\partial^2 T}{\partial Y^2}\right) \quad \text{A-15b}$$

$\frac{\partial^2 T}{\partial Z^2}$ is derived in Equation A-16.

$$\frac{\partial^2 T}{\partial Z^2} = \frac{\partial}{\partial Z}\left(\frac{\partial T}{\partial Z}\right) = \frac{\partial^2 T}{\partial Z^2} \quad \text{A-16}$$

Adding Equations A-14c, A-15b, and A-16 and also using the trigonometric relationship $cos^2\varphi + sin^2\varphi = 1$, we get Equation A-17.

$$\frac{\partial^2 T}{\partial X^2} + \frac{\partial^2 T}{\partial Y^2} + \frac{\partial^2 T}{\partial Z^2} = \frac{\partial^2 T}{\partial r^2} + \frac{1}{r}\frac{\partial T}{\partial r} + \frac{1}{r^2}\frac{\partial^2 T}{\partial \varphi^2} + \frac{\partial^2 T}{\partial Z^2} \quad \text{A-17}$$

We get the following second order partial differential Equation A-18 for heat conduction in cylindrical coordinates for constant thermal conductivity.

$$\frac{\partial^2 T}{\partial r^2} + \frac{1}{r}\frac{\partial T}{\partial r} + \frac{1}{r^2}\frac{\partial^2 T}{\partial \varphi^2} + \frac{\partial^2 T}{\partial Z^2} + \frac{\dot{q}}{k} = \frac{1}{\alpha}\frac{\partial T}{\partial \theta} \quad \text{A-18}$$

We get the following second order partial differential Equation A-19 for heat conduction in cylindrical coordinates for temperature dependent thermal conductivity.

$$\frac{\partial}{\partial r}\left(k\frac{\partial T}{\partial r}\right) + \frac{1}{r}\left(k\frac{\partial T}{\partial r}\right) + \frac{1}{r^2}\frac{\partial}{\partial \varphi}\left(k\frac{\partial T}{\partial \varphi}\right) + \frac{\partial}{\partial Z}\left(k\frac{\partial T}{\partial Z}\right) + \dot{q} = \rho C_p\frac{\partial T}{\partial \theta} \quad \text{A-19}$$

Now, let us change the second order partial differential Equation A-8 for heat conduction in Cartesian coordinates to heat conduction in spherical coordinates as shown in Figure A-3.

Cartesian coordinates are represented by spherical coordinates in Equation A-20.

$$X = rsin\psi cos\varphi;\ Y = rsin\psi sin\varphi;\ Z = rcos\psi \quad \text{A-20}$$

Partial derivatives of **X,Y,Z** with respect to **r,Φ,**ψ are given in Equations A-21.

$$\frac{\partial X}{\partial r} = sin\psi cos\varphi;\ \frac{\partial X}{\partial \varphi} = -rsin\psi sin\varphi;\ \frac{\partial X}{\partial \psi} = rcos\psi cos\varphi \quad \text{A-21a}$$

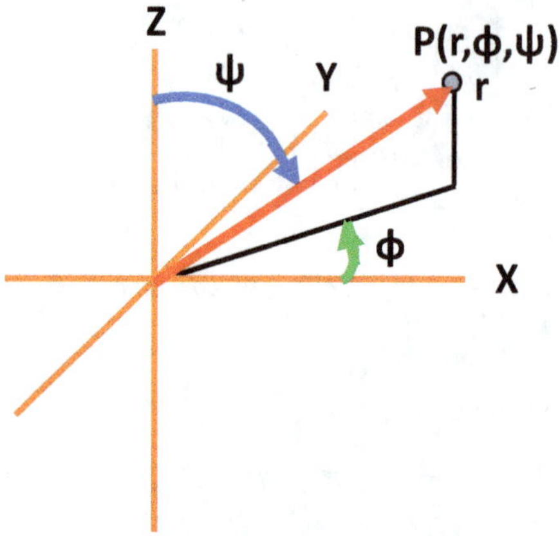

Figure A-3 Cartesian coordinates X,Y,Z and spherical coordinates r,θ,ψ . Point P(r,θ,ψ) Is defined spherical coordinates

$$\frac{\partial Y}{\partial r} = sin\psi\, sin\varphi \;;\; \frac{\partial Y}{\partial \varphi} = rsin\psi\, cos\varphi \;;\; \frac{\partial Y}{\partial \psi} = rcos\psi\, sin\varphi \qquad \text{A-21b}$$

$$\frac{\partial Z}{\partial r} = cos\psi \;;\; \frac{\partial Z}{\partial \varphi} = 0 \;;\; \frac{\partial Z}{\partial \psi} = -rsin\psi \qquad \text{A-21c}$$

We can now use the **chain rule** of differentiating because (X,Y,Z) are functions of (r,Φ,ψ) in order to obtain $\frac{\partial T}{\partial r}, \frac{\partial T}{\partial \varphi}, \frac{\partial T}{\partial \psi}, \frac{\partial^2 T}{\partial r^2}, \frac{\partial^2 T}{\partial \varphi^2}, \frac{\partial^2 T}{\partial \psi^2}$ in spherical coordinates as functions of temperature derivatives in Cartesian coordinates.

$\frac{\partial T}{\partial r}$ is shown in Equation A-22.

$$\frac{\partial T}{\partial r} = \frac{\partial T}{\partial X}\frac{\partial X}{\partial r} + \frac{\partial T}{\partial Y}\frac{\partial Y}{\partial r} + \frac{\partial T}{\partial Z}\frac{\partial Z}{\partial r} = sin\psi\, cos\varphi \frac{\partial T}{\partial X} + sin\psi\, sin\varphi \frac{\partial T}{\partial Y} + cos\psi \frac{\partial T}{\partial Z} \qquad \text{A-22}$$

$\frac{\partial T}{\partial \varphi}$ is shown in Equation A-23.

$$\frac{\partial T}{\partial \varphi} = \frac{\partial T}{\partial X}\frac{\partial X}{\partial \varphi} + \frac{\partial T}{\partial Y}\frac{\partial Y}{\partial \varphi} + \frac{\partial T}{\partial Z}\frac{\partial Z}{\partial \varphi} = -rsin\psi\, sin\varphi \frac{\partial T}{\partial X} + rsin\psi\, cos\varphi \frac{\partial T}{\partial Y} + 0 \qquad \text{A-23}$$

$\frac{\partial T}{\partial \psi}$ is shown in Equation A-24.

$$\frac{\partial T}{\partial \psi}=\frac{\partial T}{\partial X}\frac{\partial X}{\partial \psi}+\frac{\partial T}{\partial Y}\frac{\partial Y}{\partial \psi}+\frac{\partial T}{\partial Z}\frac{\partial Z}{\partial \psi}=rcos\psi cos\varphi\frac{\partial T}{\partial X}+rcos\psi sin\varphi\frac{\partial T}{\partial Y}-rsin\psi\frac{\partial T}{\partial Z} \quad \text{A-24}$$

$\frac{\partial^2 T}{\partial r^2}$ is derived in Equations A-25.

$$\frac{\partial^2 T}{\partial r^2}=sin\psi cos\varphi\frac{\partial}{\partial X}\left(\frac{\partial T}{\partial X}\right)\left(\frac{\partial X}{\partial r}\right)+sin\psi sin\varphi\frac{\partial}{\partial Y}\left(\frac{\partial T}{\partial Y}\right)\left(\frac{\partial Y}{\partial r}\right)+cos\psi\frac{\partial}{\partial Z}\left(\frac{\partial T}{\partial Z}\right)\left(\frac{\partial Z}{\partial r}\right) \quad \text{A-25a}$$

$$\frac{\partial^2 T}{\partial r^2}=sin^2\psi cos^2\varphi\frac{\partial^2 T}{\partial X^2}+sin^2\psi sin^2\varphi\frac{\partial^2 T}{\partial Y^2}+cos^2\psi\frac{\partial^2 T}{\partial Z^2} \quad \text{A-25b}$$

$\frac{\partial^2 T}{\partial \varphi^2}$ is derived in Equations A-26.

$$\frac{\partial^2 T}{\partial \varphi^2}=-rsin\psi cos\varphi\frac{\partial T}{\partial X}-rsin\psi sin\varphi\frac{\partial}{\partial X}\left(\frac{\partial T}{\partial X}\right)\left(\frac{\partial X}{\partial \varphi}\right)-rsin\psi sin\varphi\frac{\partial T}{\partial Y}+rsin\psi cos\varphi\frac{\partial}{\partial Y}\left(\frac{\partial T}{\partial Y}\right)\left(\frac{\partial Y}{\partial \varphi}\right) \quad \text{A-26a}$$

$$\frac{\partial^2 T}{\partial \varphi^2}=-rsin\psi cos\varphi\frac{\partial T}{\partial X}-r^2sin^2\psi sin^2\varphi\frac{\partial^2 T}{\partial X^2}-rsin\psi sin\varphi\frac{\partial T}{\partial Y}+r^2sin^2\psi cos^2\varphi\frac{\partial^2 T}{\partial Y^2} \quad \text{A-26b}$$

$\frac{\partial^2 T}{\partial \psi^2}$ is derived in Equations A-27.

$$\frac{\partial^2 T}{\partial \psi^2}=-rsin\psi cos\varphi\frac{\partial T}{\partial X}+rcos\psi cos\varphi\frac{\partial}{\partial X}\left(\frac{\partial T}{\partial X}\right)\left(\frac{\partial X}{\partial \psi}\right)-rsin\psi sin\varphi\frac{\partial T}{\partial Y}+$$
$$rcos\psi sin\varphi\frac{\partial}{\partial Y}\left(\frac{\partial T}{\partial Y}\right)\left(\frac{\partial Y}{\partial \psi}\right)-rcos\psi\frac{\partial T}{\partial Z}-rsin\psi\frac{\partial}{\partial Z}\left(\frac{\partial T}{\partial Z}\right)\left(\frac{\partial Z}{\partial \psi}\right) \quad \text{A-27a}$$

$$\frac{\partial^2 T}{\partial \psi^2}=-rsin\psi cos\varphi\frac{\partial T}{\partial X}+r^2cos^2\psi cos^2\varphi\frac{\partial^2 T}{\partial X^2}-rsin\psi sin\varphi\frac{\partial T}{\partial Y}+$$
$$r^2cos^2\psi sin^2\varphi\frac{\partial^2 T}{\partial Y^2}-rcos\psi\frac{\partial T}{\partial Z}+r^2sin^2\psi\frac{\partial^2 T}{\partial Z^2} \quad \text{A-27b}$$

Above partial derivatives in spherical coordinates, Equations A-22 through A-27, can be combined to show the following relationship between spherical and Cartesian coordinates as shown in Equation A-28.

$$\frac{\partial^2 T}{\partial X^2}+\frac{\partial^2 T}{\partial Y^2}+\frac{\partial^2 T}{\partial Z^2}=\frac{1}{r^2}\frac{\partial}{\partial r}\left(r^2\frac{\partial T}{\partial r}\right)+\frac{1}{r^2 sin^2\psi}\frac{\partial^2 T}{\partial \varphi^2}+\frac{1}{r^2 sin\psi}\frac{\partial}{\partial \psi}\left(sin\psi\frac{\partial T}{\partial \psi}\right) \quad \text{A-28}$$

Sum of all $\frac{\partial T}{\partial X}$ terms at the left-hand side of Equation A-28 equals to zero. Sum of all $\frac{\partial T}{\partial Y}$ terms at the left-hand side of Equation A-28 equals to zero. Sum of all $\frac{\partial T}{\partial Z}$ terms at the left-hand side of Equation A-28 equals to zero. Sum of all $\frac{\partial^2 T}{\partial X^2}$ terms at the left-hand side of Equation A-28 equals the one. Sum of all $\frac{\partial^2 T}{\partial Y^2}$ terms at the left-hand side of Equation A-28 equals the one. Sum of all $\frac{\partial^2 T}{\partial Z^2}$ terms at the left-hand side of Equation A-28 equals the one.

We get the following second order partial differential Equation A-29 for heat conduction in spherical coordinates for constant thermal conductivity.

$$\frac{1}{r^2}\frac{\partial}{\partial r}\left(r^2\frac{\partial T}{\partial r}\right)+\frac{1}{r^2 sin^2\psi}\frac{\partial^2 T}{\partial \varphi^2}+\frac{1}{r^2 sin\psi}\frac{\partial}{\partial \psi}\left(sin\psi\frac{\partial T}{\partial \psi}\right)+\frac{\dot{q}}{k}=\frac{1}{\alpha}\frac{\partial T}{\partial \theta} \quad \text{A-29}$$

We get the following second order partial differential Equation A-30 for heat conduction in cylindrical coordinates for temperature dependent thermal conductivity.

$$\frac{1}{r^2}\frac{\partial}{\partial r}\left(kr^2\frac{\partial T}{\partial r}\right)+\frac{1}{r^2 sin^2\psi}\frac{\partial}{\partial \varphi}\left(k\frac{\partial T}{\partial \varphi}\right)+\frac{1}{r^2 sin\psi}\frac{\partial}{\partial \psi}\left[sin\psi\left(k\frac{\partial T}{\partial \psi}\right)\right]+\dot{q}=\rho C_p\frac{\partial T}{\partial \theta} \quad \text{A-30}$$

Index

T

U

V

W

www.ingramcontent.com/pod-product-compliance
Lightning Source LLC
LaVergne TN
LVHW011515200926
848484LV00001B/42
9780791886786